for
space

donated by

Oliver Neumann

for
space

doreen massey

SAGE Publications
Los Angeles • London • New Delhi • Singapore

First published 2005
Reprinted 2005 , 2006 (twice), 2007

 SAGE Publications Ltd
1 Oliver's Yard
55 City Road
London EC1Y 1SP

SAGE Publications Inc.
2455 Teller Road
Thousand Oaks, California 91320

SAGE Publications India Pvt Ltd
B 1/I 1 Mohan Cooperative Industrial Area
Mathura Road, New Delhi 110 077
India

SAGE Publications Asia-Pacific Pte Ltd
33 Pekin Street #02-01
Far East Square
Singapore 048763

British Library Cataloguing in Publication data

A catalogue record for this book is available from the British Library

ISBN 978-1-4129-0361-5
ISBN 978-1-4129-0362-2 (pbk)

Library of Congress Control Number 2004094666

Typeset by C&M Digitals (P) Ltd, Chennai, India
Printed on paper from sustainable resources
Printed in Great Britain by The Cromwell Press Ltd, Trowbridge, Wiltshire

Contents

acknowledgements

This book has been written, and rewritten, over a number of years in the increasingly pressured interstices of life as 'an academic'. It would be impossible to thank everyone who has influenced my ideas, in conversations variously intense and meandering, over that time but I should like to acknowledge just a few. The Geography Department at the Open University is constantly provocative of new thoughts. Within the department John Allen, Dave Featherstone (now at Liverpool), Steve Pile, and Arun Saldanha (now at Minnesota) gave me really helpful comments on all or part of the manuscript. More widely, I gained much from seminar discussions of these ideas at a number of universities, and especially at the Geography Departments at Queen Mary, University of London, and the University of Heidelberg. An annual Reading Weekend of German-speaking geographers has been a source of inspiration and friendship. Many of the arguments here, though, have had their source, and have been tested, in the world beyond academe – in the ordinary things of life and in a whole variety of political engagements. In the process of production I have benefited from the expert help of the team at SAGE, Robert Rojek, David Mainwaring, Janey Walker and Vanessa Harwood, and from the secretarial assistance of Michele Marsh at the Open University. In particular I should like to thank Neeru Thakrar, also at the Open University, whose skills in producing the typed manuscript and her professional administrative support have been invaluable. Finally, the longest conversation has been with my sister Hilary Corton, herself by education, imagination and passion also a geographer and with whom in the course of much walking talking and general travelling many of the thoughts here have developed.

The author and publishers wish to thank the following for permission to use copyright material:

Figures

Figure 1.1a: Courtesy of The Bodleian Library, University of Oxford, MS. Arch. Selden. A. 1, fol. 2r
Figure 1.1b: Courtesy of The Newberry Library, Chicago
Figure 1.2: Courtesy of the Bibliothèque nationale de France, Paris
Figures 11.1, 12.1a and 12.2: Thanks to cartographer John Hunt at The Open University
Figure 11.2: © Tim Parfitt, www.hertfordshire.com
Figures 12.1a and 12.4: © Blackwell Publishing Ltd, Oxford
Figure 12.3: © The Palaeontological Association
Figure 13.1: Design © Steffan Böhle; used with the kind permission of Ulla Neumann
On p.140 the image is © Peter Pedley Postcards, Glossop, Derbyshire

Section opening images

Part *One*   Courtesy of The Bancroft Library, University of California, Berkeley
Part *Two*   © The MC Escher Company
Part *Three* © Steve Bell
Part *Four*   © Ann Bowker
Part *Five*   Design © Steffan Böhle; used with the kind permission of Ulla Neumann

Texts

The boxed text on p.165 is courtesy of Greenpeace (http://www.greenpeace.org)

Part *Three* develops arguments first outlined in 'Imagining Globalisation: Power-Geometries of Time-Space', Chapter 2 of *Global Futures: Migration, Environment and Globalization* edited by Atvar Brah, Mary J. Hickman and Máirtin Mac an Ghaill. Thanks to the British Sociological Association and BSA Publications Limited.

Tenochtitlan. Tierra del nopal. Entrada de Hernan Cortes, la cual se verificó el 8 de Noviembre de 1519.

Part *One*
Setting the scene

I've been thinking about 'space' for a long time. But usually I've come at it indirectly, through some other kind of engagement. The battles over globalisation, the politics of place, the question of regional inequality, the engagements with 'nature' as I walk the hills, the complexities of cities. Picking away at things that don't seem quite right. Losing political arguments because the terms don't fit what it is you're struggling to say. Finding myself in quandaries of apparently contradictory feelings. It is through these persistent ruminations – that sometimes don't seem to go anywhere and then sometimes do – that I have become convinced both that the implicit assumptions we make about space are important and that, maybe, it could be productive to think about space differently.

Three ruminations

1   The armies were approaching the city from the quarter named the reed or crocodile – the direction in which the sun rises. Much was known about them already. Tales had come back from outlying provinces. Tax gatherers from the city, collecting tribute from conquered territories, had met up with them. Envoys had been despatched, to engage in talks, to find out more. And now neighbouring groups, chafing against their long subordination to the Aztec city, had thrown in their lot with the strange invaders. Yet in spite of all these prior contacts, the constant flow of messages, rumours, interpretations reaching the city, the approaching army was still a mystery. ('The strangers sat on "deer as high as the rooftops". Their bodies were completely covered, "only their faces can be seen. They are white, as if made of lime. They have yellow hair, although some have black. Long are their beards."'[1]) And they were arriving from the geographical direction which, in these time-spaces, was held to be that of authority.

figure 1.1a   *Tenochtitlán – Aztec depiction*
Source: The Bodleian Library

It was also the Year One Reed, a year of both historical and cosmological significance: a particular point in the cycle of years. Over past cycles the city had become mightily successful. It was only a few cycles ago that the Mexica/Aztecs had first set up in this huge high valley. They had arrived from the direction of the flint and after long wanderings; an uncultivated people in the eyes of the cities already established around the lake. But since their arrival, and the founding of this city Tenochtitlán, the Aztecs had piled success upon success. The city was now the biggest in the world. Its empire now stretched, through conquest and continual violent subordination, to the ocean in two directions.

Thus far the Aztecs had conquered all before them. But these armies approaching now are ominous. Empires do not last for ever. Only recently Azcapotzalco, on the edge of the lake, had been brought down after a brief blaze of glory. And Tula, seat of the revered Toltecs, now lies deserted, as do the ruins of Teotihuacan. All these are reminders of previous splendours, and of their fragility. And now these strange invaders are coming from the direction of acatl; and it is the Year One Reed.

Such things are important. Coincidences of events form the structures of time-space. For Moctezuma they add to the whole wretched conundrum of how to respond. It could be a moment of crisis for the Empire.[2]

The men in the approaching army could hardly believe their eyes when they first looked down upon the city. They had heard that it was splendid but this was five times the size of Madrid, in the changing Europe which they had left behind just a few years ago. And these voyages, originally, had set out towards the west in the hope of finding the east. When, some years before, Cristobal Colón had 'headed across the great emptiness west of Christendom, he had accepted the challenge of legend. Terrible storms would play with his ships as if they were nutshells and hurl them into the jaws of monsters; the sea serpent, hungry for human flesh, would be lying in wait in the murky depths. … navigators spoke of strange corpses and curiously carved pieces of wood that floated in on the west wind …'[3] It was now the Year of Our Lord 1519.[4] This small army, with Hernán Cortés at its head and its few horses and its armour, had sailed from what their leaders had decided to call Cuba at the beginning of the year, and now it was November. The

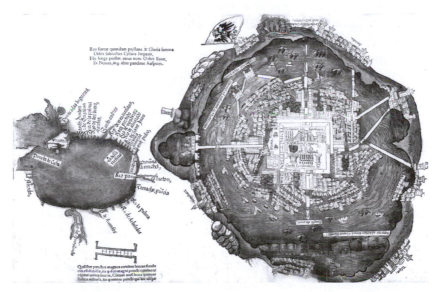

figure 1.1b   *Tenochtitlán – Spanish depiction*
Source: The Newberry Library

journey from the coast had been hard and violent, with battles and the making of alliances. Finally, now, they had heaved to the top of this pass between two snow-capped volcanoes. To Cortés' left and high above him, Popocatepetl steamed endlessly. And below him, in the distance, lay this incredible city, like nothing he had ever seen before.

There were to be two years of duplicitous negotiation, miscalculation, bloodshed, rout, retreat and readvance before Hernán Cortés, Spanish conquistador, conquered the city of the Aztecs, Tenochtitlán, which today we call la ciudad de México, Mexico City, Distrito Federal.

The way, today, we often tell that story, or any of the tales of 'voyages of discovery', is in terms of crossing and conquering space. Cortés voyaged across space, found Tenochtitlán, and took it. 'Space', in this way of telling things, is an expanse we travel across. It seems perhaps all very obvious.

But the way we imagine space has effects – as it did, each in different ways, for Moctezuma and Cortés. Conceiving of space as in the voyages of discovery, as something to be crossed and maybe conquered, has particular ramifications. Implicitly, it equates space with the land and sea, with the earth which stretches out around us. It also makes space seem like a surface; continuous and given. It differentiates: Hernán, active, a maker of history, journeys across this surface and finds Tenochtitlán upon it. It is an unthought cosmology, in the gentlest sense of that term, but it carries with it social and political effects. So easily this way of imagining space can lead us to conceive of other places, peoples, cultures simply as phenomena 'on' this surface. It is not an innocent manoeuvre, for by this means they are deprived of histories. Immobilised, they await Cortés' (or our, or global capital's) arrival. They lie there, on space, in place, without their own trajectories. Such a space makes it more difficult to see in our mind's eye the histories the Aztecs too have been living and producing. What might it mean to reorientate this imagination, to question that habit of thinking of space as a surface? If, instead, we conceive of a meeting-up of histories, what happens to our implicit imaginations of time and space?

2   The current governments in the UK and the USA (and plenty of other current governments besides) tell us a story of the inevitability of globalisation. (Or rather, although they do not of course make this distinction, they tell us a story of the inevitability of that particular form of neoliberal capitalist globalisation which we are experiencing at the moment – that duplicitous combination of the glorification of the (unequally) free movement of capital on the one hand with the firm control over the movement of labour on the other. Anyhow, they tell us it's inevitable.) And if you

point to differences around the globe, to Moçambique or Mali or Nicaragua, they will tell you such countries are just 'behind'; that eventually they will follow the path along which the capitalist West has led. In 1998 Bill Clinton delivered himself of the reflection that 'we' can no more resist the current forces of globalisation than we can resist the force of gravity. Let us pass over the possibilities of resisting the force of gravity, noting merely that this is a man who spends a good deal of his life flying about in aeroplanes …. More seriously, this proposition was delivered unto us by a man who had spent much of his recent career precisely trying to protect and promote (through GATT, the WTO, the speeding-up of NAFTA/TLC) this supposedly implacable force of nature. We know the counter argument: 'globalisation' in its current form is not the result of a law of nature (itself a phenomenon under dispute). It is a project. What statements such as Clinton's are doing is attempting to persuade us that there is no alternative. This is not a description of the world as it is so much as an image in which the world is being made.

This much is now well established in critiques of today's globalisation. But it is perhaps less often made explicit that one of the crucial manoeuvres at work within it, to convince us of the ineluctability of this globalisation, is a sleight of hand in terms of the conceptualisation of space and time. The proposition turns geography into history, space into time. And this again has social and political effects. It says that Moçambique and Nicaragua are not really different from 'us'. We are not to imagine them as having their own trajectories, their own particular histories, and the potential for their own, perhaps different, futures. They are not recognised as coeval others. They are merely at an earlier stage in the one and only narrative it is possible to tell. That cosmology of 'only one narrative' obliterates the multiplicities, the contemporaneous heterogeneities of space. It reduces simultaneous coexistence to place in the historical queue.

And so again: what if? What if we refuse to convene space into time? What if we open up the imagination of the single narrative to give space (literally) for a multiplicity of trajectories? What kinds of conceptualisation of time and space, and of their relation, might that give on to?

3   And then there is 'place'. In the context of a world which is, indeed, increasingly interconnected the notion of place (usually evoked as 'local place') has come to have totemic resonance. Its symbolic value is endlessly mobilised in political argument. For some it is the sphere of the everyday, of real and valued practices, the geographical source of meaning, vital to hold on to as 'the global' spins its ever more powerful and alienating webs. For others, a 'retreat to place' represents a protective pulling-up of drawbridges and a building of walls against the new invasions. Place, on this reading, is the locus of denial, of attempted withdrawal from

invasion/difference. It is a politically conservative haven, an essentialising (and in the end unviable) basis for a response; one that fails to address the real forces at work. It has, undoubtedly, been the background imagination for some of the worst of recent conflicts. The upheavals in 1989 in various parts of old communist Europe brought a resurgence, on a new scale and with a new intensity, of nationalisms and territorial parochialisms characterised by claims to exclusivity, by assertions of the home-grown rooted authenticity of local specificity and by a hostility to at least some designated others. But then what of the defence of place by working-class communities in the teeth of globalisation, or by aboriginal groups clinging to a last bit of land?

Place plays an ambiguous role in all of this. Horror at local exclusivities sits uneasily against support for the vulnerable struggling to defend their patch. While place is claimed, or rejected, in these arguments in a startling variety of ways, there are often shared undergirding assumptions: of place as closed, coherent, integrated as authentic, as 'home', a secure retreat; of space as somehow originarily regionalised, as always-already divided up.[5] And more than that again, they institute, implicitly but held within the very discourses that they mobilise, a counterposition, sometimes even a hostility, certainly an implicit imagination of different theoretical 'levels' (of the abstract versus the everyday, and so forth), between space on the one hand and place on the other.

What then if we refuse this imagination? What then not only of the nationalisms and parochialisms which we might gladly see thereby undermined, but also of the notion of local struggles or of the defence of place more generally? And what if we refuse that distinction, all too appealing it seems, between place (as meaningful, lived and everyday) and space (as what? the outside? the abstract? the meaning*less*)?

It is in the context of worrying away at questions such as these that the arguments here have evolved. Some of the moments that generated the thinking here I have written about before – 1989, the conflicts of class and ethnicity in east London, the elusive Frenchness of sitting in a Parisian café – but they have persisted, and crop up again here pushed a little further. Encounters with the apparently familiar but where something continues to trouble, and unexpected lines of thought slowly unwind. Most of all, the arguments which follow took shape, theoretically and politically, in the context of the perniciousness of exclusivist localisms and the grim inequalities of today's hegemonic form of globalisation; and in the face of the difficulties, too, of responding. It was wrestling with the formulation of these political issues that led to the prising open of their, often hidden, ways of conceiving of space.

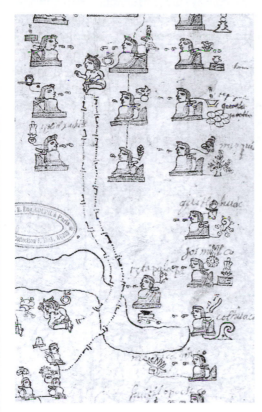

figure 1.2 *Aztec footsteps in the Codex Xolotl*

Source: Bibliothèque nationale de France

In the Year One Reed/Year of Our Lord 1519, among the many aspects of radical otherness that came face-to-face in the Valley of Mexico was the manner of imagining 'space'. Cortés carried with him aspects of an incipient version of present Western imaginations at the beginning of their triumphal progress; but imaginations still embedded in myth and emotion. For the Aztecs, too, though very differently, gods, time and space were inextricably linked. A 'basic aspect of the Aztec world view' was 'a tendency to focus on things in the process of becoming another' (Townsend, 1992, p. 122) and 'Mexica thought did not recognise an abstract space and time, separate and homogeneous dimensions, but rather concrete complexes of space and time, heterogeneous and singular sites and events. ... "place-moments" ["lugares momentos"]' (Soustelle, 1956, p. 120; my translation).

The Codex Xolotl, a hybrid construction, tells stories. Events are linked by footsteps and dotted lines between places. 'The manuscript is read by locating the origin of the footprints and deciphering the place signs as they occur on these itineraries' (Harley, 1990, p. 101). Whereas the general assumption about Western maps today is that they are representations of space, these maps, as were the European mappae mundi, were representations of time and space together.

The imagination of space as a surface on which we are placed, the turning of space into time, the sharp separation of local place from the space out there; these are all ways of taming the challenge that the inherent spatiality of the world presents. Most often, they are unthought. Those who argue that Moçambique is just 'behind' do not (presumably) do so as a consequence of much deep pondering upon the nature of, and the relationship between, space and time. Their conceptualisation of space, its reduction to a dimension for the display/representation of different moments in time, is one assumes, implicit. In that they are not alone. One of the recurring motifs in what follows is just how *little*, actually, space is thought about explicitly. None the less, the persistent associations leave a residue of effects. We develop ways of incorporating a spatiality into our ways of being in the world, modes of coping with the challenge that the enormous reality of space throws up. Produced through and embedded in practices, from quotidian negotiations to global strategising, these implicit engagements of space feed back into and sustain wider understandings of the world. The trajectories of others can be immobilised while we proceed with our own; the real challenge of the contemporaneity of others can be deflected by their relegation to a past (backward, old-fashioned, archaic); the defensive enclosures of an essentialised place seem to enable a wider disengagement, and to provide a secure foundation. In that sense, each of the earlier ruminations provides an example of some kind of failure (deliberate or not) of spatial imagination. Failure in the sense of being inadequate to face up to the challenges of space; a failure to take on board its coeval multiplicities, to accept its radical contemporaneity, to deal with its constitutive complexity. What happens if we try to let go of those, by now almost intuitive, understandings?

1

opening propositions

This book makes the case for an alternative approach to space. It has both the virtue, and all the disadvantages, of appearing obvious. Yet the ruminations above, and much that is to come, imply that it still needs elaborating.

It is easiest to begin by boiling it down to a few propositions. They are the following. *First*, that we recognise space as the product of interrelations; as constituted through interactions, from the immensity of the global to the intimately tiny. (This is a proposition which will come as no surprise at all to those who have been reading recent anglophone geographical literature.) *Second*, that we understand space as the sphere of the possibility of the existence of multiplicity in the sense of contemporaneous plurality; as the sphere in which distinct trajectories coexist; as the sphere therefore of coexisting heterogeneity. Without space, no multiplicity; without multiplicity, no space. If space is indeed the product of interrelations, then it must be predicated upon the existence of plurality. Multiplicity and space as co-constitutive. *Third*, that we recognise space as always under construction. Precisely because space on this reading is a product of relations-between, relations which are necessarily embedded material practices which have to be carried out, it is always in the process of being made. It is never finished; never closed. Perhaps we could imagine space as a simultaneity of stories-so-far.

Now, these propositions resonate with recent shifts in certain quarters in the way in which progressive politics can also be imagined. Indeed it is part of my argument, not just that the spatial is political (which, after many years and much writing thereupon, can be taken as given), but rather that thinking the spatial in a particular way can shake up the manner in which certain political questions are formulated, can contribute to political arguments already under way, and – most deeply – can be an essential element in the imaginative structure which enables in the first place an opening up to the very sphere of the political. Some of these possibilities can already be drawn out from the brief statement of propositions. Thus, although it would be incorrect, and too rigidly constraining, to propose any simple one-to-one mapping, it is possible to elucidate from each a slightly different aspect of the potential range of connections between the imagination of the spatial and the imagination of the political.

Thus, *first*, understanding space as a product of interrelations chimes well with the emergence over recent years of a politics which attempts a commitment to anti-essentialism. In place of an individualistic liberalism or a kind of identity politics which takes those identities as already, and for ever, constituted, and argues for the rights of, or claims to equality for, those already-constituted identities, this politics takes the constitution of the identities themselves and the relations through which they are constructed to be one of the central stakes of the political. 'Relations' here, then, are understood as embedded practices. Rather than accepting and working with already-constituted entities/identities, this politics lays its stress upon the relational constructedness of things (including those things called political subjectivities and political constituencies). It is wary therefore about claims to authenticity based in notions of unchanging identity. Instead, it proposes a relational understanding of the world, and a politics which responds to that.

The politics of interrelations mirrors, then, the first proposition, that space too is a product of interrelations. Space does not exist prior to identities/entities and their relations. More generally I would argue that identities/entities, the relations 'between' them, and the spatiality which is part of them, are all co-constitutive. Chantal Mouffe (1993, 1995), in particular, has written of how we might conceptualise the relational construction of political subjectivities. For her, identities and interrelations are constituted together. But spatiality may also be from the beginning integral to the constitution of those identities themselves, including political subjectivities. Moreover, specifically spatial identities (places, nations) can equally be reconceptualised in relational terms. Questions of the geographies of relations, and of the geographies of the necessity of their negotiation (in the widest sense of that term) run through the book. If no space/place is a coherent seamless authenticity then one issue which is raised is the question of its internal negotiation. And if identities, both specifically spatial and otherwise, are indeed constructed relationally then that poses the question of the geography of those relations of construction. It raises questions of the politics of those geographies and of our relationship to and responsibility for them; and it raises, conversely and perhaps less expectedly, the potential geographies of our social responsibility.

Second, imagining space as the sphere of the possibility of the existence of multiplicity resonates with the greater emphasis which has over recent years in political discourses of the left been laid on 'difference' and heterogeneity. The most evident form which this has taken has been the insistence that the story of the world cannot be told (nor its geography elaborated) as the story of 'the West' alone nor as the story of, for instance, that classic figure (ironically frequently itself essentialised) of the white, heterosexual male; that these were particular stories among many (and that their understanding through the eyes of the West or the straight male is itself specific). Such trajectories were part of a complexity and not the universals which they have for so long proposed themselves to be.

The relationship between this aspect of a changing politics (and manner of doing social theory) and the second proposition about space is of a rather different nature from in the case of the first proposition. In this case, the argument is that the very possibility of any serious recognition of multiplicity and heterogeneity itself depends on a recognition of spatiality. The political corollary is that a genuine, thorough, spatialisation of social theory and political thinking can force into the imagination a fuller recognition of the simultaneous coexistence of others with their own trajectories and their own stories to tell. The imagination of globalisation as a historical queue does not recognise the simultaneous coexistence of other histories with characteristics that are distinct (which does not imply unconnected) and futures which potentially may be so too.

Third, imagining space as always in process, as never a closed system, resonates with an increasingly vocal insistence within political discourses on the genuine openness of the future. It is an insistence founded in an attempt to escape the inexorability which so frequently characterises the grand narratives related by modernity. The frameworks of Progress, of Development and of Modernisation, and the succession of modes of production elaborated within Marxism, all propose scenarios in which the general directions of history, including the future, are known. However much it may be necessary to fight to bring them about, to engage in struggles for their achievement, there was always none the less a background conviction about the direction in which history was moving. Many today reject such a formulation and argue instead for a radical openness of the future, whether they argue it through radical democracy (for example Laclau, 1990; Laclau and Mouffe, 2001), through notions of active experimentation (as in Deleuze and Guattari, 1988; Deleuze and Parnet, 1987) or through certain approaches within queer theory (see as one instance Haver, 1997). Indeed, as Laclau in particular would most strongly argue, only if we conceive of the future as open can we seriously accept or engage in any genuine notion of politics. Only if the future is open is there any ground for a politics which can make a difference.

Now, here again – as in the case of the first proposition – there is a parallel with the conceptualisation of space. Not only history but also space is open.[6] In this open interactional space there are always connections yet to be made, juxtapositions yet to flower into interaction (or not, for not all potential connections have to be established), relations which may or may not be accomplished. Here, then, space is indeed a product of relations (first proposition) and for that to be so there must be multiplicity (second proposition). However, these are not the relations of a coherent, closed system within which, as they say, everything is (already) related to everything else. Space can never be that completed simultaneity in which all interconnections have been established, and in which everywhere is already linked with everywhere else. A space, then, which is neither a container for always-already constituted identities nor a completed closure of holism. This is a space of loose ends and missing links. *For the future to be open, space must be open too.*

All these words come trailing clouds of connotations. To write of challenging the opposition between space and place might legitimately provoke thoughts of Heidegger (but that is not what I mean). Talking of 'difference' can engender assumptions about othering (but that is not what I am getting at). Mention of multiplicities evokes, among others, Bergson, Deleuze, Guattari (and there will be some engagement later with that strand of thought). A few preliminary clarifications might help.

By 'trajectory' and 'story' I mean simply to emphasise the process of change in a phenomenon. The terms are thus temporal in their stress, though, I would argue, their necessary spatiality (the positioning in relation to other trajectories or stories, for instance) is inseparable from and intrinsic to their character. The phenomenon in question may be a living thing, a scientific attitude, a collectivity, a social convention, a geological formation. Both 'trajectory' and 'story' have other connotations which are not intended here. 'Trajectory' is a term that figures in debates about representation that have had important and abiding influences on the concepts of space and time (see the discussion in Part *Two*). 'Story' brings with it connotations of something told, of an interpreted history; but what I intend is simply the history, change, movement, of things themselves.

That bundle of words difference/heterogeneity/multiplicity/plurality has also provoked much contention. All I mean at this point is the contemporaneous existence of a plurality of trajectories; a simultaneity of stories-so-far. Thus the minimum difference occasioned by being positioned raises already the fact of uniqueness. This is, then, not 'difference' as opposed to class, as in some old political battles. It is simply the principle of coexisting heterogeneity. It is not the particular nature of heterogeneities but the fact of them that is intrinsic to space. Indeed it puts into question what might *be* the pertinent lines of differentiation in any particular situation. Nor is this 'difference' as in the deconstructive move of spac*ing*: as in the deconstruction of discourses of authenticity, for instance. This does not mean that such discourses are not significant in the cultural moulding of space; nor that they should not be taken to task. Romances of coherent nationhood, as in the third rumination, may operate on precisely such principles of constituting identity/difference. David Sibley (1995, 1999) among others has explored such attempts at the purification of space. Indeed, they are precisely one way of coping with its heterogeneities – its actual complexity and openness. But the point at issue here is another one: not negative difference but positive heterogeneity. This links back to the political argument against essentialism. Insofar as that argument adopted a form of social constructionism which was confined to the discursive, it did not in itself offer a positive alternative. Thus in the particular case of space, it may help us to expose some of its presumed coherences but it does not properly bring it to life. It is that liveliness, the complexity and openness of the configurational itself, the positive multiplicity, which is important for an appreciation of the spatial.

This book is an essay on the challenge of space, the multiple ruses through which that challenge has been so persistently evaded, and the political implications of practising it differently. In pursuit of this there is inevitable engagement with many other theorists and theoretical approaches, including many whose explicit focus is not always on spatiality. They are referenced in the text. But it is perhaps important to say now that my argument is not simply in the mould of any one of them. I have not worked from texts on space but through situations and engagements in which the question of space has in some way been entangled. Rather, my preoccupation with pushing away at space/politics has moulded positions *on* philosophy, and on a range of concepts. The debates about heterogeneity/difference and social constructionism/discourse are cases in point. Equations of representation with spatialisation have troubled me; associations of space with synchrony exasperated me; persistent assumptions of space as the opposite of time have kept me thinking; analyses that remained within the discursive have just not been positive enough. It has been a reciprocal engagement. What I'm interested in is how we might imagine spaces for these times; how we might pursue an alternative imagination. What is needed, I think, is to uproot 'space' from that constellation of concepts in which it has so unquestioningly so often been embedded (stasis; closure; representation) and to settle it among another set of ideas (heterogeneity; relationality; coevalness … liveliness indeed) where it releases a more challenging political landscape.

There has, as is often now recounted, been a long history of understanding space as 'the dead, the fixed' in Foucault's famous retrospection. More recently and in total contrast there has been a veritable extravaganza of non-Euclidean, black-holey, Riemannian … and a variety of other previously topologically improbable evocations. Somewhere between these two lie the arguments I want to make. What you will find here is an attempt to awaken space from the long sleep engendered by the inattention of the past but one which remains perhaps more prosaic, though none the less challenging, than some recent formulations. That is what I found to be most productive. This is a book about ordinary space; the space and places through which, in the negotiation of relations within multiplicities, the social is constructed. It is in that sense a modest proposal, and yet the very persistence, the apparent obviousness, of other mobilisations of 'space', point to its continuing necessity.

There are many who have pondered the challenges and delights of *temporality*. Sometimes this has been done through the lens of that strand of anthropocentric

13

philosophical miserabilism which preoccupies itself with the inevitability of death. In other guises temporality has been extolled as the vital dimension of life, of existence itself. The argument here is that space is equally lively and equally challenging, and that, far from it being dead and fixed, the very enormity of its challenges has meant that the strategies for taming it have been many, varied and persistent.

When I was a child I used to play a game, spinning a globe or flicking through an atlas and jabbing down my finger without looking where. If it landed on land I'd try to imagine what was going on 'there' 'then'. How people lived, the landscape, what time of day it was, what season. My knowledge was extremely rudimentary but I was completely fascinated by the fact that all these things were *going on now*, while I was here in Manchester in bed. Even now, each morning when the paper comes, I cast my eye down at the world's weather (100˚F and cloudy in New Delhi, 46 and raining in Santiago; 82 and sunny in Algiers). It's partly a way of imagining how things are for friends in other places; but it's also a continuing amazement at the contemporaneous heterogeneity of the planet. (I wrote this book under the working title of 'Spatial delight'.) It was, possibly still is, all appallingly naive, and I have learned at least some of its dangers. The grotesqueness of the maps of power through which aspects of this 'variety' can be constituted; the real problems of thinking about, and still more of appreciating, place; how much more easy it is for some than for others to forget the simultaneity of those different stories; the difficulty simply, even, of travelling. (The telling of the voyages of discovery in a way that holds 'the discovered' still; the version of globalisation which dismisses others to the past …) None the less it seems important to hold on to an appreciation of that simultaneity of stories. It sometimes seems that in the gadarene rush to abandon the singularity of the modernist grand narrative (the singular universal story) what has been adopted in its place is a vision of an instantaneity of interconnections. But this is to replace a single history with no history – hence the complaint, in this guise, of depthlessness. In this guise, the 'spatial turn' were better refused. Rather we should, could, replace the single history with many. And this is where space comes in. In that guise, it seems to me, it is quite reasonable to take some delight in the possibilities it opens up.

Part *Two* addresses some of the imaginations of space that we inherit from a range of philosophical discourses. This is not a book about philosophy but at this point it engages *with* some strands of philosophy in order to argue that from them are derived common readings and associations which may help to explain why in social and political life we so often lend to space the characteristics we do. Part *Three* takes up a range of ways in which space is articulated in social theory and in practical-popular and political engagements, in particular in the context of debates about modernity and capitalist globalisation. In neither of these Parts is the primary aim one of critique: it is to pull out the positive threads which enable a more lively appreciation of the challenge of space. Part *Four* then elaborates a range of further reorientations concerning both space and place. Throughout the book, strands of the relevance of these arguments to political debate are developed, and Part *Five* turns to these directly. This book, then, is not 'for space' in preference to something else; rather it is an argument for the recognition of particular characteristics of space and for a politics that can respond to them.

A number of subthemes weave their way *sotto voce* through the Parts. Some of these have their own headings. The series called 'A reliance on science?' questions some elements of the current relation between natural and social sciences broadly conceived. 'The geography of knowledge production' weaves a story of the connection between certain modes of practising science and the social and geographical structures in which they are set (indeed, more strongly, through which they are constituted). In both of these spheres, it is proposed, not only are there implicit spatialities but also there are both conceptual and political links to the wider argument of the book.

Other themes persistently surface as part of the more general thesis. There is an attempt to go beyond the specifically human. There is a commitment to the old theme that space matters, but also a questioning of some of the ways in which it is commonly thought to do so. There is an attempt to work towards a groundedness that – in an age in which globalisation is so easily imagined as some kind of force emanating always from 'elsewhere' – is vital for posing political questions. There is an insistence, relatedly, on specificity, and on a world neither composed of atomistic individuals nor closed into an always-already completed holism. It is a world being made, through relations, and there lies the politics. Finally, there is an urge towards 'outwardlookingness', towards a positivity and aliveness to the world beyond one's own turf, whether that be one's self, one's city, or the particular parts of the planet in which one lives and works: a commitment to that radical contemporaneity which is the condition of, and condition for, spatiality.

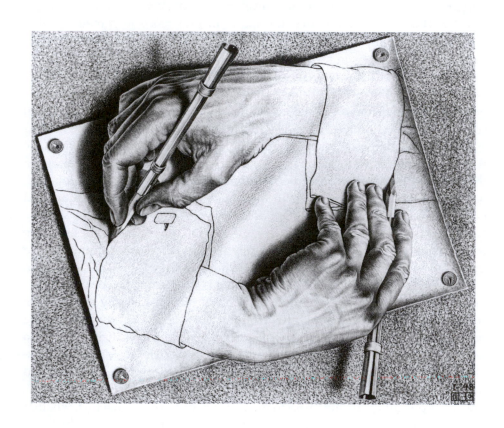

Part *Two*
Unpromising associations

Henri Lefebvre points out in the opening arguments of *The production of space* (1991) that we often use that word 'space', in popular discourse or in academic, without being fully conscious of what we mean by it. We have inherited an imagination so deeply ingrained that it is often not actively thought. Based on assumptions no longer recognised as such, it is an imagination with the implacable force of the patently obvious. That is the trouble.

That implicit imagination is fed by all kinds of influences. In many cases they are, I want to argue, unpromising associations which connotationally deprive space of its most challenging characteristics. The influences to be addressed in this Part derive from philosophical writings in the broadest sense of that term. Part *Three* will take up more practical-popular and social-theoretical understandings of space, particularly in the context of the politics of modernity and capitalist globalisation. The aim of both Parts is to unearth some of the influences on hegemonic imaginations of 'space'. What follows immediately, then, is an attempt to draw out some particular threads of argument which exemplify ways in which space can come, through significant philosophical discourses, to have associated with it characteristics which, to my mind at least, disable its full insertion into the political. This is not a book about philosophy; the arguments here are particular and focus solely on how some commonly accepted positions, even if not directly concerned with space, have reverberations none the less for the way in which we imagine it. The particular philosophical strands addressed here serve as exemplars. They revolve around Henri Bergson, structuralism and deconstruction: a selection made both because of their significance as strands of thought and because in their wider arguments they have, in different ways, much to offer the kind of project this book is engaged in. In other words, they are engaged with because of their promise rather than their problems.

None of these philosophers has the reconceptualisation of space as their objective. Most often, and in the context of wider debates, temporality is a more pressing concern. Over and again space is conceptualised as (or, rather, assumed to be) simply the negative opposite of time. It is indeed, I want to argue, in part that lacuna in relation to thinking actively about space, and the contradictions which thereby arise, that can provide a hint of how to breach the apparent limits of some of the arguments as they now stand. One theme is that time and space must be thought together: that this is not some mere rhetorical flourish, but that it influences how we think of both terms; that thinking of time and space together does not mean they are identical (for instance in some undifferentiated four-dimensionality), rather it means that the imagination of one will have repercussions (not always followed through) for the imagination of the other and that space and time are implicated in each other; that it opens up some problems which have heretofore seemed (logically, intractably) insoluble; and that it has reverberations for thinking about politics and the spatial. Thinking about history and temporality necessarily has implications (whether we recognise them or not) for how we imagine the spatial. The counterpositional labelling of phenomena as temporal or spatial, and entailing all the baggage of the reduction of space to the a-political sphere of causal closure or the reactionary redoubts of established power, continues to this day.

The prime aims of the philosophies explored here were largely in tune with the arguments presented in this book. I cheer on Bergson in his arguments about time, approve of structuralism's determination not to let geography be turned into history, applaud Laclau's insistence on the intimate connection between dislocation and the possibility of politics … It's just when they get to talking about space that I find myself rebuffed. Puzzled by the lack of explicit attention they give, irritated by their assumptions, confused by a kind of double usage (where space is both the great 'out there' and the term of choice for characterisations of representation, or of ideological closure), and, finally, pleased sometimes to find the loose ends (their own internal dislocations) which make possible the unravelling of those assumptions and double usages and which, in turn, provokes a reimagination of space which might be not just more to my liking, but also more in tune with the spirit of their own enquiries.

There is one distinction which ought to be made from the outset. It has been argued that, at least in recent centuries, space has been held in less esteem, and has been accorded less attention, than has time (within geography, Ed Soja (1989) has made this argument with force). It is often termed the 'prioritisation of time over space' and it has been remarked on and taken to task by many. It is not, however, my concern here. What I am concerned with is the *way* we imagine space. Sometimes the problematical character of this imagination does indeed perhaps result from deprioritisation – the conceptualisation of space as an afterthought, as a residual of time. Yet the early structuralist thinkers can by no means be said to have prioritised time and still, or so I shall argue, the effect of their approach was a highly problematical imagination of space.

Moreover, the excavation of these problematical conceptualisations of space (as static, closed, immobile, as the opposite of time) brings to light other sets of connections, to science, writing and representation, to issues of subjectivity and its conception, in all of which implicit imaginations of space have played an important role. And these entwinings are in turn related to the fact that space has so often been excluded from, or inadequately conceptualised in relation to, and has thereby debilitated our conceptions of, politics and the political.

What follows is an engagement with some of those debilitating associations. Each of these strands of philosophy has developed in particular historico-geographical conjunctures. They themselves have been interventions in something already moving. Sometimes what is at issue is disentangling them in some measure from the orientations provoked by their moments, the debates of which they were a part. Reorienting them to my own concerns can produce new lines of thought from them. Sometimes what is at issue is pushing them further. The effect in the end, I hope, is to liberate 'space' from some chains of meaning (which embed it with *closure* and *stasis*, or with *science*, *writing* and *representation*) and which have all but choked it to death, in order to set it into other chains (in this chapter alongside *openness*, and *heterogeneity*, and *liveliness*) where it can have a new and more productive life.

2
space/representation

There is an idea with such a long and illustrious history that it has come to acquire the status of an unquestioned nostrum: this is the idea that there is an association between the spatial and the fixation of meaning. Representation – indeed conceptualisation – has been conceived of as spatialisation. The various authors who will figure in this chapter have come to this position along different routes, but almost all of them subscribe to it. Moreover, though the reference is to 'spatialisation', there is in all cases slippage; it is not just that representation is equated with spatialisation but that the characteristics thus derived have come to be attributed to space itself. Moreover, though the further development of these philosophical positions implies almost always quite another understanding of what space might be, none of them pause very long either explicitly to develop this alternative or to explore the curious fact that this other (and more mobile, flexible, open, lively) view of space stands in such flat opposition to their equally certain association of representation with space. It is an old association; over and over we tame the spatial into the textual and the conceptual; into representation.

Of course, the argument is usually quite the opposite: that through representation we spatialise time. It is space which is said thereby to tame the temporal.

Henri Bergson's is one of the most complex and definitive of these philosophical positions. For him, the burning concern was with temporality, with 'duration'; with a commitment to the experience of time and to resisting the evisceration of its internal continuity, flow and movement. It is an attitude which strikes chords today. In *Bergsonism*, Deleuze (1988) denounces what he sees as our exclusive preoccupation with extended magnitudes at the expense of intensities. As Boundas (1996, p. 85) expands this, the impatience is with our over-insistent focus on the discrete at the expense of continua, things at the expense of processes, recognition at the expense of encounter, results at the expense of tendencies … (and lots more besides). Every argument being proposed in this book would support such an endeavour. A reimagination of things as processes is necessary (and indeed now widely accepted) for the reconceptualisation of places in a way that might challenge exclusivist localisms based on claims of some eternal authenticity. Instead of things as pregiven discrete entities, there is now a move towards recognising the continuous becoming which is in the nature of their being. Newness, then, and creativity, is an essential characteristic of temporality. And in *Time and free will* (1910), Bergson plunges straight into an engagement with psychophysics and the science of his day, wielding an argument that this intellectualisation was taking the life out of experience. By conceptualising, by dividing it up, by writing it down, it was obliterating that vital element of life itself.

To address the problem he worked through a distinction between different kinds of multiplicities. For both Bergson and Deleuze, whom Boundas (1996) rolls together, in relation to this discussion, as Deleuze–Bergson, are engaged over the meanings of 'difference' and 'multiplicity'. For them there is an important distinction between *discrete* difference/multiplicity (which refers to extended magnitudes and distinct entities, the realm of diversity) and *continuous* difference/multiplicity (which refers to intensities, and to evolution rather than succession). The former is divided up, a dimension of separation; the latter is a continuum, a multiplicity of fusion. Both Bergson and Deleuze are in battle to instate the significance, indeed the philosophical primacy, of the second (continuous) form of difference over the first (the discrete) form. What is at issue is an insistence on the genuine openness of history, of the future. For Bergson, change (which he equated with temporality) implies real novelty, the production of the really new, of things not already totally determined by the current arrangement of forces. Once again, then, there is a real coincidence of desires with the argument of this book. For the burden of the third proposition of this book is precisely to argue not just for a notion of 'becoming', but for the openness of that process of becoming.

However, Bergson's overwhelming concern with time, and his desire to argue for its openness, turned out to have devastating consequences for the way he conceptualised space. This has often been attributed to a classic (modernist?) prioritisation of time. Indeed Soja (1989) argues that Bergson was one of the most forceful instigators of a more general devaluation and subordination of space relative to time which took place during the second half of the nineteenth century (see also Gross, 1981–2). And the classic recantation by Foucault of the long history of the denigration of space, begins: 'Did it start with Bergson, or before?' (Foucault, 1980, p. 70). The problem however runs more deeply than simple prioritisation. Rather, it is a question of the mode of conceptualisation. It is not so much that Bergson 'deprioritised' space, as that *in the association of it with representation* it was deprived of dynamism, and radically counterposed to time. Thus:

> Has true duration anything to do with space? Certainly, our analysis of the idea of number [which he has just been discussing] could not but make us doubt this analogy, to say no more. For if time, as the reflective consciousness represents it, is a medium in which our conscious states form a discrete series so as to admit of being counted, and if on the other hand our conception of number ends in spreading out in space everything which can be directly counted, it is to be

presumed that time, understood in the sense of a medium in which we make distinctions and count, is nothing but space. That which goes to confirm this opinion is that we are compelled to borrow from space the images by which we describe what the reflective consciousness feels about time and even about succession; it follows that pure duration must be something different. Such are the questions which we have been led to ask by the very analysis of the notion of discrete multiplicity. But we cannot throw any light upon them except by a direct study of the ideas of space and time in their mutual relations. (1910, p. 91)

One of the crucial provocations for Bergson, and a constant reference point, is Zeno's paradox. The message which the paradox is used to hammer home is that movement (a continuum) cannot be broken up into discrete instants. 'It is … because the continuum cannot be reduced to an aggregate of points that movement cannot be reduced to what is static. Continua and movements imply one another' (Boundas, 1996, p. 84). This is an important argument but it is an argument about the nature of *time*, about the impossibility of reducing real movement/becoming to stasis multiplied by infinity; the impossibility of deriving history from a succession of slices through time (see also Massey, 1997a).

However the line of thought gets tangled up with an idea (inadvertent? certainly not very explicit) of space. Thus, in *Matter and Memory* (Bergson, 1911) we find:

> The arguments of Zeno of Elea have no other origin than this illusion. They all consist in making time and movement coincide with the line which underlies them, in attributing to them the same subdivisions as to the line, in short in treating them like that line. In this confusion Zeno was encouraged by common sense, which usually carries over to the movement the properties of its trajectory, and also by language, which always translates movement and duration in terms of space. (p. 250)

The rejected time of instantaneous time-slices attracts the label 'spatial', as in: what is at stake for Bergson–Deleuze is 'the primacy of the heterogeneous time of [temporal] difference over the spatialized time of metrication with its quantitative segments and instants' (Boundas, 1996, p. 92). Immediately this association renders space in a negative light (as the lack of 'movement and duration'). And so, to the list of dualisms within which these philosophies are doing combat (continua rather than discontinuities, processes rather than things…) is added time rather than space (p. 85).

Now these arguments have taken flight in particular situations. One dragon that had to be vanquished (but which is still around today) was empty time. Empty, divided and reversible time in which nothing changes; where there is no evolution but merely succession; a time of a multiplicity of discrete things. Bergson's concern was that time is too often conceptualised in the same manner as space (as a discrete multiplicity). We misunderstand the nature of duration, he argued, when we 'spatialize' it – when we think of it as a fourth dimension of extension. (There is here a prescient critique of an over-easy tendency to talk of space-time, or of four-dimensionality, without investigating the nature of the integration of dimensions which is at issue.) The nature of the dragon provoked the form of the response. The instantaneous slice through time was assumed to be static, as it is in the form in which it is invoked in Zeno's paradox. It was then awarded the label 'spatial'. And finally it was argued: anyway, if there is to be real becoming (the genuine continuous production of the new), then such sup-posedly static slices through time must be impossible. Static time-slices, even multiplied to infinity, cannot produce becoming.

However, the argument can be turned around. Does not the argument in the form just recounted imply that the 'space' which comes to be defined, *via* a con-notational connection with representation, must likewise be impossible? Does it not rather mean that space itself (the dimension of a discrete multiplicity) can precisely *not* be a static slice through time? With that kind of space it would indeed be impossible to have history as becoming. In other words, not only can time *not* be sliced up (transforming it from a continuous to a discrete multi-plicity) but even the argument that this is *not* possible should not refer to the result as space. The slide here from spatialisation as an activity to space as a dimension is crucial. Representation is seen to take on aspects of spatial*isation* in the latter's action of setting things down side by side; of laying them out as a discrete simultaneity. But representation is also in this argument understood as fixing things, taking the time out of them. The equation of spatialisation with the production of 'space' thus lends to space not only the character of a discrete multiplicity but also the characteristic of stasis.

Space, then, is characterised as the dimension of quantitative divisibility (see, for instance, *Matter and Memory*, 1911, pp. 246–53). This is fundamental to the notion that representation is spatialisation: 'Movement visibly consists in pass-ing from one point to another, and consequently in traversing space. Now the space which is traversed is infinitely divisible; and as the movement is, so to speak, applied to the line along which it passes, it appears to be one with this line and, like it, divisible' (p. 248). This character of space as the dimension of plurality, discrete multiplicity, is important, both conceptually and politically. But in Bergson's formulation here it is a discrete multiplicity *without duration*. It is not only instantaneous it is static. Thus, 'we cannot make movement out of immobilities, nor time out of space' (*Time and Free Will*, 1910, p. 115). From a number of angles, this proposition will be questioned in the argument which follows. In *Matter and Memory* Bergson writes 'The fundamental illusion consists in transferring to duration itself, in its continuous flow, the form of the instanta-neous sections which we make in it' (1911, p. 193). In its intent I applaud this argument; but I would demur at its terms. Why can we not imbue these instan-taneous sections with their own vital quality of duration? A dynamic simul-taneity would be a conception quite different from a frozen instant (Massey, 1992a). (And then, if we persisted in the nomenclature of 'spatial' we could indeed 'make time out of space' – save that we would not have started from such a counterpositional definition in the first place.) On the one hand, this throws doubt upon the use of the word 'space' in the foregoing quotations from Bergson; on the other hand, however, it shows that the very impetus of his argument provides a further step, a questioning of the use of the term space itself. It is a questioning already implicit in Bergson's argument, even in these earlier works.

The problem is that the connotational characterisation of space through representation, as not only discrete but also without life, has proved strong. Thus, Gross (1981–2) writes of Bergson as arguing that 'the rational mind merely spatialises', and that he conceptualised scientific activity in terms of 'the immobilising (spatial) categories of the intellect':

> For Bergson, the mind is by definition spatially oriented. But everything creative, expansive and teeming with energy is *not*. Hence, the intellect can never help us reach what is essential because it kills and fragments all that it touches … We must, Bergson concluded, break out of the spatialisation imposed by mind in order to regain contact with the core of the truly living, which subsists only in the time dimension … (pp. 62, 66; emphasis in the original)

As Deleuze (1988) persistently points out, this is to load the cards. Space and time here are not two equal but opposing tendencies; everything is stacked on the side of duration. This 'principal Bergsonian division: that between duration and space' (p. 31) provides its own way forward through its very imbalance. 'In Bergsonism, the difficulty seems to disappear. For by dividing the composite according to two tendencies, with only one showing the way in which a thing varies qualitatively in time, Bergson effectively gives himself the means of choosing the "right side" in each case' (p. 32).

In *Creative evolution* (Bergson, 1911/1975), the distinction between spatialisation and space is made effective. While retaining the equation between intellectualisation and spatialisation ('The more consciousness is intellectualized, the more is matter spatialized', p. 207), Bergson came to recognise also, at first in the form of a question, the duration in external things and this in turn pointed to a radical change in the potential conceptualisation of space. That recognition of the duration in external things and thus the interpenetration, though not the equivalence, of space and time is an important aspect of the argument in this book. It is what I am calling space as the dimension of multiple trajectories, a simultaneity of stories-so-far. Space as the dimension of a multiplicity of durations. The problem has been that the old chain of meaning – space–representation–stasis – continues to wield its power. The legacy lingers on.

Thus, for Ernesto Laclau (1990) the development of the argument is rather different from Bergson's but the conclusion is similar: 'space' is equivalent to representation which in turn is equivalent to ideological closure.[1] For Laclau spatialisation is equivalent to hegemonisation: the production of an ideological closure, a picture of the essentially dislocated world as somehow coherent. Thus:

> any representation of a dislocation involves its spatialization. The way to overcome the temporal, traumatic and unrepresentable nature of dislocation is to construct it as a moment in permanent structural relation with other moments, in which case the pure temporality of the 'event' is eliminated ... this spatial domesticization of time ... (p. 72)[2]

Laclau equates 'the crisis of all spatiality' (as a result of the assertion of dislocation's constitutive nature) with 'the ultimate impossibility of all representation' (p. 78) ... 'dislocation destroys all space and, as a result, the very possibility of representation' (p. 79), and so on. The pointers towards a potential reformulation are evident and exciting (if all space is destroyed...?), but they are not followed up, and the assumption of an equivalence between space and representation is unequivocal and insisted-upon.

In contrast yet again to Laclau, who rather tends just to *assume* that representation is spatialisation, de Certeau, who holds the same position, spells out in some detail his reasons why. They are very similar to Bergson's. For de Certeau, the emergence of writing (as distinct from orality) and of modern scientific method involved precisely the obliteration of temporal dynamic, the creation of a blank space (*un espace propre*) both of the object of knowledge and as a place for inscription, and the act of writing (on that space). These three processes are intimately associated. Narratives, stories, trajectories are all suppressed in the emergence of science as the writing of the world. And that process of writing, more generally of making a mark upon the blank space of a page, is what removes the dynamism of 'real life'. Thus, in his attempt, which is really the whole burden of his book, to invent ways of recapturing those narratives and stories (precisely to bring them back into some form of produced 'knowledge') he ruminates upon whether or not to use the word 'trajectory'. The term, he thinks,

> suggests a movement, but it also involves a plane projection, a flattening out. It is a transcription. A graph (which the eye can master) is substituted for an operation; a line which can be reversed (i.e. read in both directions) does duty for an irreversible temporal series, a tracing for acts. To avoid this reduction, I resort to a distinction between *tactics* and *strategies*. (de Certeau, 1984, p. xviii–xix; emphasis in the original)

Now, this association of scientific writing with assumptions of reversibility, and a desire to hang out for irreversibility, harks back to the engagements which Bergson had with the science of his day. Science-writing takes the life out of processes, and renders them reversible; whereas real life is irreversible. A first reflection on this will be explored later: that we should no longer be fighting that battle against 'science' – both because Science is not a source of unimpugnable truth (though it is most certainly a powerful discourse), and because there are now plenty of scientists who would anyway no longer hold this position.

De Certeau continues:

> However useful this 'flattening out' may be, it transforms the *temporal* articulation of places into a *spatial* sequence of points. (p. 35; emphasis in the original)

Moreover, the distinction de Certeau makes is once again related directly and explicitly to representation:

> … the occasion – that indiscreet instant, that poison – has been controlled by the spatialization of [i.e. by] scientific discourse. As the constitution of a proper place, scientific writing ceaselessly reduces time, that fugitive element, to the normality of an observable and readable system. In this way, surprises are averted. Proper maintenance of the place eliminates these criminal tricks. (p. 89)

And finally he writes of:

> … the (voracious) property that the geographical system has of being able to transform action into legibility, but in doing so it causes a way of being in the world to be forgotten. (p. 97)

Ironically, it is on the basis of this argument that de Certeau decides against the use of the term 'trajectory' and instead resorts to a distinction between tactics and strategy which cements into place precisely the dualism (including between space and time) with which the rest of the book is struggling.[3]

One way and another, then, all of these authors equate space and representation. It is a remarkably pervasive and unquestioned assumption, and it does indeed have an intuitive obviousness. But as already indicated perhaps this equation of representation and spatialisation is *not* something which should be taken for granted. At the very least its implacability and its repercussions might be disturbed. It is an extraordinarily important move. For what it does is to associate the spatial with stabilisation. Guilt by association. Spatial layout as a way of containing the temporal – both its terrors and its creative delights. Spatialisation, on this view, flattens the life out of time. I want, through the course of this book, to build an argument which will come to a very different conclusion.

To begin with, note that there are two things going on here: first, the argument that representation necessarily fixes, and therefore deadens and detracts from, the flow of life; and second, that the product of this process of deadening is space. The first proposition I would not entirely dispute, although the form in which it is customarily couched is presently being modified. However, it seems to me that there is no case at all for the second proposition: that there is an equivalence between space and representation. It is one of those accepted things that are by now so deeply embedded that they are rarely if ever questioned. Let us, then, question it.

In order to ground the discussion, it is necessary to establish some preliminary points.

First, it is important in itself to recognise that this way of thinking has a history. It derives, as do all positions, from social embeddedness and intellectual/ scientific engagement. From the very earliest days of Western philosophy the capturing of time in a sequence of numbers has been thought of as its spatialisation. The appeal of this has already been acknowledged. The problem lies in the movement from spatialisation to characterisations of space. Citations tracing the persistence of that imagination could be numerous, and tedious. Perhaps just one, to give the essence of the case: Whitehead (1927/1985) writes of 'the presentational immediacy' of space which 'enables space to speak for the less accessible dimension of time, with differences in space being used as a surrogate for differences in time' (pp. 21–3). I shall suggest that one route of development for this now-hegemonic equation of space and representation may thread its way through nineteenth-century and early twentieth-century battles over the meaning of time. This is not, of course, in any way to 'criticise': such embeddedness is inevitable. It is merely to emphasise that this intellectual position is the product of a process: it is not somehow self-evident.

Second, even if we agree that representation indeed fixes and stabilises (though see below), what it so stabilises is not simply time, but space-time. Laclau writes of *'history's* ultimate unrepresentability' (1990, p. 84; my emphasis), but what is really unrepresentable is not history conceived of as temporality but time-space (history/geography if you like). Indeed, two pages earlier he both half-recognises this (by referring to 'society') but then blows it by his use of space-terminology: 'Society, then, is ultimately unrepresentable: any representation – and thus any space – is an attempt to constitute society, not to state what it is' (p. 82). It would be better to recognise that 'society' is both temporal and spatial, and to drop entirely that definition of representation as space. What is at issue, in the production of representations, is not the spatialisation of time (understood as the rendering of time as space), but the representation of time-space. What we conceptualise (divide up into organs, put it how you will) is not just time but space-time. In the arguments of Bergson and de Certeau too the issue is formulated as though the lively world which is there to be represented (conceptualised/written down) is only temporal. It certainly *is* temporal; but it is spatial too. And 'representation' is an attempt to capture both aspects of that world.

Third, it is easy to see how representation can be understood as a form of spatialisation. That business of laying things out side by side; indeed the production of a simultaneity, a discrete multiplicity. (On this basis space would also be easy to represent, if that were merely what space was.) So Bergson writes of substituting the path for the journey, de Certeau of substituting a tracing for acts. But consider. In de Certeau's formulation, a tracing is itself a representation; it is not 'space'. The map is not the territory. Alternatively, what Bergson writes is: 'You substitute the path for the journey, and because the journey is subtended by the path you think the two coincide' (1911, p. 248). We may, here, though it is set within a wider discussion of representation, take the path to be a real path (not a representation/conceptualisation). It is not the map; it is the territory itself. But then a territory is integrally spatio-temporal. The path is not a static instantaneity. Indeed, we can now draw out Laclau's own conclusions. All space, he writes as we have seen, is dislocated. A first consequence is Laclau's own point: that there is a crisis of representation (in the sense that it must be recognised as constitutive rather than mimetic). But a second consequence is that space itself, the space of the world, far from being equivalent to representation, must be *un*representable in that latter, mimetic, sense.

This historically significant way of imagining space/spatialisation not only derives from an assumption that space is to be defined as a lack of temporality (holding time still) but also has contributed substantially to its continuing to be thought of in that way. It has reinforced the imagination of the spatial as petrification and as a safe haven from the temporal, and – in the images which it almost inevitably invokes of the flat horizontality of the page – it further makes 'self-evident' the notion of space as a surface. All these imaginaries not only diminish our understanding of spatiality but, through that, they even make more difficult the project which was central to all of these authors: that of opening up temporality itself.

Now, there have in recent years been challenges both to representation as any kind of 'mirror of nature' (Rorty, 1979; and many others) and as an attempt to de-temporalise. On the latter, Deleuze and Guattari, for instance, argue that a concept should express an event, a happening, rather than a de-temporalised essence and (drawing indeed on Bergson) argue against any notion of a tripartite division between reality, representation and subjectivity. Here what we might have called representation is no longer a process of fixing, but an element in a continuous production; a part of it all, and itself constantly becoming. This is a position which rejects a strict separation between world and text and which understands scientific activity as being just that – an activity, a practice, an embedded engagement *in* the world of which it is a part. Not representation but experimentation. It is an argument which has been made by many (for instance Ingold, 1993; Thrift, 1996) across a range of disciplines. Together with the notion of the text/representation as itself an open disseminatory network, it at least begins to question the understanding of scientific practice as representation-as-stabilisation in that sense. The geographers Natter and Jones (1993) trace parallels between the histories of representation and space, suggesting that the post-structuralist critique of representation-as-mirror could be re-enacted as a parallel critique of space. As the text has been destabilised in literary theory so space might be destabilised in geography (and indeed in wider social theory).

The issue is complex, however. For if scientific/intellectual activity is indeed to be understood as an active and productive engagement in/of the world it is none the less a *particular kind* of practice, a specific form of engagement/ production in which it is hard to deny (to absolve ourselves from the responsibility for?) *any* element of representation (see also Latour, 1999b; Stengers, 1997), even if it is, quite certainly, productive and experimental rather than simply mimetic, and an embodied knowledge rather than a mediation. It does not, however, have to be conceived of as producing a space, nor its characteristics carried over to inflect our implicit imaginations of space. For to do so is to rob space of those characteristics of freedom (Bergson), dislocation (Laclau) and surprise (de Certeau) which are essential to open it up to the political.

It is peculiar that space is so widely imagined as 'conquering time'. It seems in general to be perceived that space is somehow a lesser dimension than time: one with less gravitas and magnificence, it is the material/phenomenal rather than the abstract; it is being rather than becoming and so forth; and it is feminine rather than masculine (see, for instance, Bondi, 1990; Massey, 1992a; Rose, 1993). It is the subordinated category, almost the residual category, the not-A to time's A, counterpositionally defined simply by a lack of temporality, and widely seen as, within modernity, having suffered from deprioritisation in relation to time.

And yet this denigrated dimension is so often seen as conquering time. For Laclau, 'Through dislocation time is overcome by space. But while we can speak of the hegemonization of time by space (through repetition), it must be emphasized that the opposite is not possible: time cannot hegemonize anything, since it is a pure effect of dislocation' (1990, p. 42). For de Certeau, 'the "proper" is a victory of space over time' (1984, p. xix). The victory is of course one of 'representation' over 'reality', of stabilisation over life, where space is equated with representation and stabilisation (and therefore time, one is forced to presume, with reality and life). The language of victory reinforces an imagination of enmity between the two. But life is spatial as well as temporal. Walker (1993), writing of international relations theory, argues that 'modern accounts of history and temporality have been guided by attempts to capture the passing moment within a spatial order' (pp. 4–5). He points to that 'fixing of temporality within spatial categories that has been so crucial in the construction of the most influential traditions of Western philosophy and socio-political thought' (p. 4). Likewise in anthropology Fabian (1983) has developed at length an argument that a core, and debilitating, assumption of that discipline has been its spatialisation of time: 'the temporal discourse of anthropology as it was formed decisively under the paradigm of evolutionism rested on a conception of Time that was not only secularized and naturalized but also thoroughly spatialized' (p. 16).

Thus the supposedly weaker term of a dualism obliterates the positive characteristics of the stronger one, the privileged signifier. And it does this through the conflation of the spatial with representation. Space conquers time by being set up as the *representation of* history/life/the real world. On this reading space is an order imposed upon the inherent life of the real. (Spatial) order obliterates (temporal) dislocation. Spatial immobility quietens temporal becoming. It is, though, the most dismal of pyrrhic victories. For in the very moment of its conquering triumph 'space' is reduced to stasis. The very life, and certainly the politics, are taken out of it.

(A reliance on science? 1)

Sotto voce *through much of that story of the connotational connection of representation with space has run another thread: that of the relationship between this connection and conceptualisations of 'science'.*

The most evident relationship is where 'science' stands for the whole process of representation (the trace rather than the journey), and thus in fact for intellectual knowledge in general. The whole business of conceptualisation; the intellectual rather than the lived or the intuitive.

But the engagement with science was also more immediately and specifically with the natural sciences. Bergson's practice, in particular, had deep roots in the historical development of the natural sciences and in their complex relationship with philosophy. Time and free will *plunges straight in as Bergson does battle with the ascendant psychophysics of his day. It is clearly that which has provoked him, motivated him into his present argument. And there were other wrestlings, too, with Riemann over the nature of multiplicities, and most famously over the implications of the new relativity theory. In other words, the definition of space was caught up in the broader dialogue between the 'natural' and 'human' sciences. That was one of the encounters through which 'space' became sedimented into a particular chain of meanings. It is true once again today: people reach to the natural sciences in their efforts to conceptualise the new spaces of our times. Bergson's story, however, points to some of the difficulties of that strategy.*

Bergson's concern was with the nature of time; through 'duration' he was emphasising its continuity, its irreversibility, its openness. However, as Prigogine and Stengers (1984) document, the development of science (and in particular physics) from Newton through to and including Einstein and (some versions of) quantum mechanics operates with a notion of reversible time. Processes are reversible and there is no meaningful distinction between past and future. There have been arguments, both within science and between 'science' (in that specific form) and its doubters, but the notion of the non-reversibility of time was a hard one to establish. Timeless processes do not generate a notion of open historical time. Behind that powerful model of 'science' as 'physics in the guise of classical mechanics' is an assumption about time that deprives it of its openness; reduces its possibility of being truly historical. This is the case not only in the concept of fully timeless processes, but also in closed equilibrium systems, where the future is given, contained within the initial conditions – it is closed.

While this was accepted by many within philosophy (and indeed this form of physics, as classical mechanics, was widely adopted as a model for science – and even knowledge – in general) there were other strands of philosophy which struggled against it.[4] *'Science's' vision flew in the face of what these critical philosophers understood of* the world. A long history of the development of ideas about time (and thus, as a by-product, implicit or explicit, about space) was set in train.

The question inevitably arose of how to reconcile Science's view of the world (as static, recurring, a-temporal) with the apparently plain fact of human experience of the difference between past and future, of a very distinct, and irreversible, temporality. The hard sciences were obdurate. As Prigogine and Stengers write, the difficulty of getting 'science' to recognise an irreversible temporality 'led to discouragement and to the feeling that, in the end, the whole concept of irreversibility has a subjective origin' (1984, p. 16). 'That kind' of temporality, in other words, if it doesn't exist in Nature, must be a product of human consciousness (ignore for the minute the dualisms in play here – they were part of what constituted the blockage that had to be overcome). As Prigogine and Stengers put it, at that historical moment the choice seemed to be either to accept the pronouncements of classical science or to resort to a metaphysical philosophy based on the human experiential production of time. Both Bergson and Whitehead, among others, according to Prigogine and Stengers took the latter route. And thus there developed a whole discourse around the 'philosophy of time' which stood on the ground of individual experience. (Some of the problems must have been evident: Whose human mind are we talking about here? What kind of human mind? And how can we reconcile it anyway with what 'science' was saying about the world? But at this point in the dialogue between science and other thinkers maybe there seemed no other way out.) Bergson, it is important to say again, was subsequently to broaden his position and to argue that temporal irreversibility is fundamental to the order of things themselves.

There was, however, another question. For these 'nomad' philosophers were not interested only in some formal distinction between past and future. Rather, as we have seen, what was crucial was that the future must be open, must be there to be made. Thus, concepts of equilibrium, developed in the context of closed isolated systems, may have a notion of 'time' in them in the sense that things happen, but it is a time, a change, (a future), which is already given in the initial conditions.[5] It is not a genuinely open future of possibilities, of creation. It was precisely in trying to struggle free of such constraints that Bergson wrote 'time is invention or nothing at all' (1959, p. 784) and that Whitehead argued that there was a creativity in nature 'whereby the actual world has its character of temporal passage to novelty' (1978, no page number, cited in Prigogine, 1997, p. 59). What was at issue in these engagements was not just a need to account for 'human experience' but also a determination not to submit to determinism. The argument was about keeping history open.

Perhaps, therefore, we might understand some of the philosophical preoccupation with time, and the nature of that preoccupation, as being at least in part bound up with the struggle over the meaning of classical science. Maybe the misreading of space, its relegation to the outer darkness of fixity and closure, came about in part because of social scientists' and philosophers' reactions to natural science's intransigence on the matter of time. It was as a result of science's intransigence that some philosophers sought a way around its propositions. If time was to be asserted as open and creative, then that business that science got up to, pinning things down (writing them down) and taking the life out of them, must be its opposite – which they called 'space'.

The evolution of this story-line is indeed the burden of much of Prigogine and Stengers' book Order out of chaos. *But what Prigogine and Stengers do not do is to draw out the ramifications of this history for the conceptualisation of space. Through Western knowledge-systems, they argue, runs a dichotomy. In one corner classical science with its commitment to time-reversibility, to determinism, to the (supposed) stasis of Being. In the other corner, social science and philosophy engaging in notions of temporality, probability and the indeterminism of Becoming. However, what Prigogine and Stengers also argue is that (some of) natural science is now changing (or, at least, that it must now change) its own view of time: that new reconceptualisations of physics lead towards the recognition of an open and fully historical notion of time. So natural science itself must change, and is indeed beginning to do so: 'The results of nonequilibrium thermodynamics are close to the views expressed by Bergson and Whitehead. Nature is indeed related to the creation of unpredictable novelty, where the possible is richer than the real' (Prigogine, 1997, p. 72).*

This latter view is now recited to the point of tedium. My point here is that its history has implications for the question which Prigogine and Stengers do not take up – the one about space. For what their reading of new developments in natural sciences means, is that the science against which Bergson and others constructed their ideas no longer has to be combated: 'the limitations Bergson criticized are beginning to be overcome, not by abandoning the scientific approach or abstract thinking but by perceiving the limitations of the concepts of classical dynamics and by discovering new formulations valid in more general situations' (Prigogine and Stengers, 1984, p. 93). This must also mean that, insofar as it was influenced by the battle it was waging at the time, some of the impetus for Bergson's own earlier formulations has now dissolved.

To begin with, there may be no need to assert the irreversibility and openness of time through recourse to some idealisation of human subjectivity (see also Grosz, 2001). As Prigogine puts it, 'Figuratively speaking, matter at equilibrium is "blind", but with the arrow of time, it begins to "see". Without this new coherence due to irreversible, non-equilibrium processes, life on earth would be impossible to envision. The claim that the arrow of time is "only phenomenological", or subjective, is therefore absurd' (1997, p. 3). Indeed, not only is it absurd it is impossible, for '[i]f the world were formed by stable dynamical systems, it would be radically different from the one we observe around us. It would be a static, predictable world, but we would not be here to make the predictions' (1997, p. 55). Most significantly at this point, however: the implication is that we are not obliged to follow the conclusions of this line of argument which relate to space.

Henri Bergson was a 'nomad' in his day, part of what is now hailed as 'an orphan line of thinkers', which includes Lucretius, Hume, Spinoza, Nietzsche and Bergson and on which Deleuze has powerfully drawn (Massumi, 1988, p. x).[6] But some of the debates in relation to which Bergson ranged his arguments have now shifted, or are shifting. Today it seems that in his engagement with the dominant science as it then was, the very dynamics of his nomadism served to generate thoughts which were unfortunately to confine the conceptualisation of space.

That story of Bergson's engagement with science, and the wider debates both within philosophy and between natural scientists and a range of critical philosophers, is full of pointers for today. Bergson's was a real engagement with those sciences: aware, critical, argumentative, as well as constructively adding to them, providing ontological counterparts (Deleuze, 1988). Today again debates about space (among many other things) are frequently infused with references to natural science and to mathematics. Sometimes this is again an intervention, a proposal about the direction of science (Deleuze may be seen in this light). Often, though, it is not now a questioning relationship, nor one which takes seriously the new imaginations emerging from those sciences, to debate with them or to add to them, as Bergson did. Rather, now, the dominant tendency seems to be to borrow imaginations (fine) but also to claim their legitimacy through references to natural science. On what basis, now, do the social sciences and humanities so casually and so frequently litter their writings with references to fractals, to quanta and to complexity theory?

The frustration of Bergson, and of other philosophers, derived not only from the specifics of what natural scientists were arguing about time, but also from the emerging role and status of those sciences and especially of physics within the conventions and practice of knowledge production as a whole. In the long history stemming from Newtonian mechanics there has developed a mutual commitment and admiration between science-as-physics and philosophy-as-positivism/analytical philosophy. Such philosophy, for which all single titles seem hopelessly inadequate but which was immensely powerful in its reverberating effects, expecially in its early days and in the writings of people such as Carnap (1937), maintained that 'science' was the only road to knowledge and that there was only one true scientific method. It committed itself to (its understandings of) objectivity, the empirical method and epistemological monism (which essentially incorporated a reductionism-to-physics). The story is well known. In spite of subsequent debates, and later writings such as those of Kuhn, this relation of mutual admiration is still powerful.

And it has led both to an imagined hierarchy among the sciences (with physics at one end and, say, cultural studies and humanities at the other) and to a phenomenon of physics envy among a range of scientific practices which aim to ape, but find they cannot, the protocols of physics. Physical geographers (sometimes) think they are more 'scientific' than human geographers.[7] Neoclassical economics has striven to distinguish itself from other social sciences, to give itself as much as possible the appearance of a 'hard' science (the consequences of this in limiting its potential as a form of knowledge would be comical were they not, in their effects through analysis and policy, so tragic). Geologists suffer from physics envy: 'the sense of inferiority concerning the status of geology as compared with other, "harder" sciences …' (Frodeman, 1995, p. 961; see also Simpson, 1963). And so do biologists: 'a sense of inferiority, of "physics envy" (which may perhaps be why these days many molecular biologists try to behave as if they are physicists!)' (Rose, 1997, p. 9). It is an envy that is deeply embedded. And it still, including in our ways of conceptualising space, goes on.

Yet the Bergson story, set in an era of the establishment of physics' pomp, also points to some of the reasons why this notion of a hierarchy of sciences might be challenged.

Most evidently, the established status of physics, of its methodology and its truth claims, is based on an image of that discipline that is now out of date. Physics itself has been changing. The physics of which Prigogine writes, along with many other branches of that discipline, do not fit that Newtonian-mechanics-derived model at all.[8]

Moreover, with the benefit of being able to look back at the Bergson story with a little historical distance, what intrigues is that some of the most serious questions about openness, the nature of history and the conceptualisation of time, were being raised by philosophers. Natural scientists, on the whole, dug in their heels, ruled the questions out of court. Physics is not always 'in the lead'; we cannot appeal to it for some grounding for other (merely social, merely human) theories (Stengers, 1997). In the Bergson story maybe natural science could with benefit have listened to and learned from philosophy and social science. Thus Elizabeth Grosz, in exploring a similar theme, has written that:

Bergson ... frequently remarked on the subordination of temporality to spatiality, and consequently the scientific misrepresentation of duration. Time has been represented in literature and poetry more frequently and ably than in science. Questions about mutability and eternity are raised in philosophical speculation long before they were addressed scientifically, their stimulus coming from theology as much as from mechanics. (Grosz, 1995, p. 98)

One could cite a multitude of examples. Kroeber understands the poet Shelley confronting, and accepting, randomness and openness, in a way in which 'the most enlightened science of Shelley's day', which 'was still basically mechanistic', could not even approach (Kroeber, 1994, pp. 106–7). Mazis sees 'science' catching up with philosopher Merleau-Ponty: 'This sense of a world, made up of open systems interacting as self-ordering phenomena within a temporal flow, brings science to an ontology like that articulated by Merleau-Ponty' (1999, p. 232). As Deleuze (1995) has it, the influences can flow both ways and 'no special status should be assigned to any particular field, whether philosophy, science, art, or literature' (p. 30). Hayles (1999) makes the same argument about the relationship between science and literature. The whole business of the relationship between natural and human sciences must be understood historically, not as a one-way flow of true science to lesser practices of knowledge production, but as an exchange, a complicated, difficult, but definitely multi-directional, relationship.

All of this disturbs the ground of some of social science's current and highly contradictory relationship to the natural sciences. References to the natural sciences cannot be mobilised as some kind of final corroboration, nor as resort to a higher court whose forms of knowledge production give them an authority to which on occasions it is convenient to appeal. In the era of classical science, and on the issue of time, social science and philosophy were clearly reaching for questions which the dominant natural scientists of their day simply did not grasp. Moreover (and in case you were tempted to point to an inconsistency here) my citing of Prigogine (Nobel Prize winner in a natural science, etc.) is not done in the manner of reference to the unimpugnable authority of 'science', for there are as many fierce debates amongst natural scientists about these matters as there are amongst philosophers and social scientists. Rather, it is simply to demonstrate that, on this subject of time (and therefore I would argue, space), we no longer have to battle against 'a science' which appears monolithically to say the opposite.

3

the prison-house of synchrony

Through many twentieth-century debates in philosophy and social theory runs the idea that spatial framing is a way of containing the temporal. For a moment, you hold the world still. And in this moment you can analyse its structure.

You hold the world still in order to look at it in cross-section. It seems a small, and perhaps even an intuitively obvious, gesture, yet it has a multitude of resonances and implications. It connects with ideas of structure and system, of distance and the all-seeing eye, of totality and completeness, of the relation between synchrony and space. And – or so I want to argue – the assumptions which may lie within it and the logics to which it can give rise run off in a whole range of problematical directions.

The 'spaces' of structuralism

It is, perhaps, through the development of structuralism that we can see some of these arguments most clearly. The aim of structuralism in fact seemed to be to put space, rather than time, on to the intellectual agenda. Structuralists were involved in different intellectual contests, and were attempting to combat different enemies, from those addressed by Bergson. While for the latter the engagement was with natural science, for structuralist anthropologists the contest was with the dominance of narrative. In part this was motivated by a desire to escape the conceptualisation of certain other societies (the kind anthropologists tended then to study) as simply forebears of that of the West; as, for instance, 'primitive'. Structuralism was in part an attempt to escape precisely that convening of geography into history (though they didn't think of it quite like that) which was exemplified by the second rumination in Part *One*. The aim, an aim with which the argument of this book would totally agree, was to escape from turning world geography into a historical narrative. To achieve that aim they insisted on the coherence of each society as a structure in its own right.

In an attempt to escape the assumption of cause in narrativity, and of progression from the savage to the civilised, structuralism turned to the concepts of structure, space and synchrony. Instead of narrative, structure; instead of diachrony, synchrony; instead of time, space. It was a move made with the best of intentions. And yet, in relation to space – the very thing it was supposed to be foregrounding – it has left a legacy of assumptions and taken-for-granted understandings which have continued to this day to bedevil debate.

For what happened was that this reconceptualisation was translated (I would say *mis*translated) into notions of time and space. The structuralists were arguing against the dominance of narrativity, which was interpreted as temporality (diachrony, etc., etc.). And in their eagerness to do this (to argue against an assumed dominance of temporality) they equated their a-temporal structures with space. If these structures weren't temporal, they must be spatial. Structure and process were read as space and time. Space was conceived (or perhaps this is too active a verb – it was simply *assumed*) to be the absolute negation of time.

This is immediately evident in the easy elision between sets of terms. Thus these 'structures', being devised in order to examine the synchronic and being 'therefore' characterised by an absence of the temporal (itself a formulation which is problematical and to which we shall return), was blessed with the nomenclature of the spatial. In the great debates between the likes of Lévi-Strauss, Sartre, Braudel and Ricoeur, that counterposition of elisions (or chains of virtually equivalent meaning), between narrative/temporality/diachrony on the one hand and structure/spatiality/synchrony on the other, came to be embedded as a formulation shared between two otherwise-antagonistic positions. If they couldn't agree about anything else, they agreed about this. Or at least, which comes to the same thing, they didn't discuss it. They simply, silently, shared it. In geography, Soja among others took up the idea, writing that structuralism had been 'one of the twentieth-century's most important avenues for the reassertion of space in critical social theory' (Soja, 1989, p. 18). It is easy to see the attractions of this view. It seems to offer the opportunity to see everything all together, to understand the interconnections rather than the dynamics which push on the narrative flow. It is perhaps the 'rather than' which foreshadows the problems to come.[9]

This way, indeed, lie dangers. To begin with, although the structuralists' structures may be synchronic there is little in their definition to say that they are spaces. The argument in some ways parallels that about representation. The 'synchronic structures' of the structuralists were analytical schema devised for understanding a society, myth, or language. Structuralism goes further, then, than simply 'holding the world still'. It is quite different from 'a slice through time'. As Osborne puts it, synchrony must be distinguished from the instant. 'Synchrony is not con-temporality, but a-temporality' (1995, p. 27). Moreover, the (implicit) reason that these analytical structures are dubbed spatial is precisely that they are established as a-temporal, as the opposite of temporality *and therefore* without time, *and therefore* space. It is, primarily, a negative definition. In the logic of this reasoning space is assumed to be both the opposite of time and without temporality. Once again, although through a completely different route from that followed by Bergson, and ironically one with the avowed intention of prioritising spatiality, space is rendered as the sphere of stasis and fixity. It is a conceptualisation of space which, once again, is really a residualisation and derives from an assumption: that space is opposed to time and lacking in temporality. Thought of like this, 'space' really would be the realm of closure and that in turn would render it the realm of the impossibility of the new and therefore of the political.

Fabian (1983) argues trenchantly that Lévi-Strauss is anyway actually somewhat dissimulating in his use of the term 'space'. In his elaboration of this, Fabian brings out many confusions which are important to the argument here and by no means specific to Lévi-Strauss. 'His ruse', writes Fabian, 'is to substitute diachrony for history. That sleight of hand is supported, much like the diversions all illusionists try to create while operating their magic, by directing the reader's attention to something else, in this case to the "opposition" of Space and Time' (p. 54). Moreover, he argues, 'Lévi-Strauss leads us to believe that *space* here could mean real space, perhaps the space of the human geographers' (emphasis in the original) … while it is actually a taxonomic space, indeed a map. 'Real space', in other words, is confused once again with representation. And once again the confusion has had spectacular ramifications for our (implicit) imaginations *of* that space. In this case, however, they work not through concerns about the spatialisation of time in a discrete multiplicity (the trace for the journey) but rather through an imagination of the spatial as a synchronic *closure*. This happens in a number of ways.

First, such structures rob the objects to which they refer of their inherent dynamism. They do indeed try to 'hold the world still' but this eliminates also any possibility of real change. Osborne, though still oddly deploying the nomenclature of space, describes it well: 'a purely analytical space in which the temporality immanent to the objects of inquiry is repressed' (1995, pp. 27–8). It is a conceptual schema which is anyway lacking, and this was, of course, not a problem which went unrecognised. Lévi-Strauss himself was ambivalent about the relationship of his structures to stasis and dynamism. It was evidently undeniable that the world moves and changes. Yet what structuralism famously made of this was a conceptualisation of the world in terms of an invariant model on the one hand and variable history on the other. Jakobson (1985) insisted upon the 'interplay of invariants and variations' (p. 85); and the classic distinction between *langue* and *parole* is of the same nature. The problem such an initiating conceptualisation poses, of course, is how the two terms of the binary can be related. And the recurrent response (by no means confined to structuralism) has been to invent a third term which must have the magical properties of carrying one safely over the impasse. The resulting rickety 'solution' has been called 'ternary': it has three elements – (i) the synchronic element, (ii) the diachronic or contingent historical aspect and (iii) the bridge between the other two (Lechte, 1994). Lévi-Strauss, finding himself in a corner with only the first two terms to hand, indeed argued that the presence of a third element is always necessary (Lévi-Strauss, 1945/1972, 1956/1972). Such a third term, clearly, in order effectively to do the necessary business, has to have power-ful yet also malleable properties. It was thus that *mana* was mobilised in Lévi-Strauss' work, and myth, and facial painting among the Caduveo Indians. It is a strategy with a long history; Plato's concept of *chora* in the *Timaeus* is a simi-lar device in an attempt to cross an unbridgeable chasm. The problem as ever lies in the founding conceptualisation. And it is a founding binary conceptual-isation which has done much to mould our imaginations of what is space and what is time and how they are (supposedly) opposed. While time is history (in various forms), space is regarded as the stasis of a synchronic structure. This is just the first of many ramifications of this approach for the way in which we conceptualise the spatial.

For, *second*, the structures of the structuralists have a further feature, in addi-tion to their presumed spatiality. They are closed.[10] If there is a sense in which their definition as spatial could be said to entail a positive conceptualisation of space (rather than a negative definition as spatial because they are a-temporal) then it is because what they are concerned with is relations between coexisting elements or terms. They are about *relations*. And one of the potential implica-tions of this is that not only might we productively conceptualise space in terms of relations but also relations can only be fully recognised by thinking fully spatially. In order for there to be relations there must of necessity be spacing. However, the conceptual synchronies of structuralism are relations imagined in a highly particular way. Above all, they are characterised by relations between their constituent elements such that they form a completely interlocked system. They are closed systems. It is this aspect of the conceptualisation – in combina-tion with a-temporality – which does most damage. For the stasis of closed sys-tems robs 'relational construction' of the anti-essentialism to which it is often claimed to lead. And the closure itself robs 'the spatial' (when it is called such) of one of its potentially disruptive characteristics: precisely its juxtaposition, its happenstance arrangement-in-relation-to-each-other, of previously uncon-nected narratives/temporalities; its openness and its condition of always being made. It is this crucial characteristic of 'the spatial' which constitutes it as one of the vital moments in the production of those dislocations which are necessary to the existence of the political (and indeed the temporal). But that is to run ahead.

The legacy of structuralism lingers on. Indeed, it is more active than this. Many of its framing conceptualisations continue to influence the shape of intellectual arguments today, through from the work of Louis Althusser to the most recent engagements within post-structuralism.

There are many who still wrestle either implicitly or explicitly with the structuralists' notion of synchrony. What is striking is how the basic terms of the counterposition (temporality/a-temporality) and its elision with time/space are so frequently maintained.

Althusser attacked both the structuralist notion of synchrony and the Hegelian concept of 'essential section'. In effect, he criticised both the 'longitudinal' and the 'cross-sectional' characteristics of the Hegelian notion of historical time (see 1970, p. 94). On the one hand he took issue with the homogeneous temporality that is so essential to the Hegelian way of thinking. Althusser, like Lévi-Strauss in fact, was after a more complex understanding of history which could allow for the possibility of (indeed, in the Althusserian formulation, which *assumed*) the coexistence of different temporalities. On the other hand he took issue with the 'contemporaneity' of the Hegelian cross-section. There were two aspects to this latter point. The first concerned the relationship between parts and whole. For Althusser one of the most serious problems with Hegel's formulation was its character of being 'an expressive totality, i.e., a totality all of whose parts are so many *"total parts"*, each expressing the others, and each expressing the social totality that contains them, because each in itself contains in the immediate form of its expression the essence of the totality itself' (1970, p. 94; emphasis in the original). The potential repressiveness inherent in such a way of viewing society and the difficulty of thinking real difference, let alone 'alterity', is evident. Althusser also produced a second critique, however, which although clearly related to the first has distinct and significant implications. This is that the Hegelian essential section is characterised by total instantaneous interconnectivity: 'all the elements of the whole revealed by this section are in an immediate relationship with one another, a relationship that immediately expresses their internal essence' (p. 94). As Althusser argues, and as subsequent writers have frequently underlined (e.g. Young, 1990), the combined effect of these characteristics is to provide the necessary basis for the assumption of a singular universal. It is a notion of time, and of cross-sections through time (which are frequently called 'space'), which does not allow for really 'other' voices. This is thus a fundamentally political element of the critique. Here space cannot be the sphere of the possibility of real heterogeneity. The totally interconnected configuration both assumes a homogeneous temporality and is a prerequisite for any proposition of a singular universal.

Now, once again, the explicit focus of this debate was time. Althusser did not, explicitly, relate his critique to concepts of space; his concern was rather with thinking through the possible nature of disrupted temporalities. And yet the implications for understanding spatiality are significant. Abandoning the notion of spatiality implicit in the whole viewpoint of essential sections opens up the possibility of thinking space in an alternative way, and with interruptive and dislocating consequences. It is precisely this total-interlock which robs the structure (and thus 'the spatial' when it is characterised as such) of one of its most disruptive characteristics – its enablement of new relations-to-each-other of previously disparate trajectories. There is moreover a further line of argument which has the potential for unearthing equally political implications. The notion of a section in which all the elements exist in an immediate relationship with one another is essentially a description of a closed system. It is a system – once again – in which all the specified relationships are within the section and where all the elements in the section are tied in. It is therefore, for both of these reasons, a mode of conceptualisation which implies an inherent stasis of the cross-section. And in so far as the cross-section, to distinguish it from the temporality of the longitudinal story, becomes characterised as 'spatial', such a mode of conceptualisation reduces space to precisely that causally closed sphere of the nothing-doing that robs it of all political potential and which was referred to above in the discussion of structuralism.

Although some commentators (e.g. Osborne, 1995, p. 27) express surprise, Althusser was therefore quite right to criticise the structuralists for adopting these aspects of the Hegelian section in their concept of 'synchrony'. Where Althusser was mistaken was in equating the Hegelian essential section with the structuralists' synchrony (Osborne (p. 27), also makes this point).[11] The two are not the same. While the former may be more easily equated to the temporal instant, the latter is the no-time of the causally closed system. It is a-temporal in a double sense: in that it is a conceptual formulation *un-related* to time; and in that in its causal closure it disallows real change, and therefore politics. Indeed, the more fundamental problem, as Althusser recognised, is the whole notion of counterposition between synchrony and diachrony. If synchronies are causally closed, then the diachronic can be no more than a sequence of synchronies. This characteristic they do indeed share with the Hegelian essential section. On all these readings 'history' turns out to be a-historical; it is reduced to a series of slices through time. Merely a series of 'spaces', internally interconnected cross-sections, following each other in sequence.

Althusser's work, then, points to two rather different intellectual sources for this particular imagination of space as a dimension which is the opposite of time, and as a dimension lacking in temporality. On the one hand there are Hegelian notions of a single totalised history within which, at every moment – which is of necessity a moment of total contemporaneity – every part is expressive of the whole. On the other hand there is the legacy of dubbing as space the a-temporal structures/synchronies of the structuralists. Both have political implications. Space has been read by many as a-political because it is conceptualised as a seamless whole; as the totally interconnected closed system of a synchronic structure. It is not dislocated, and 'dislocation is the source of freedom' (Laclau, 1990, p. 60).

It is lacking in the contingency which is the condition for that openness which, in turn, is the precondition for politics.[12] Moreover, that view of the coherence of space in turn enables the existence of only one history, one voice, one speaking position. The inheritance, for the spatial, has thus been glum. Space has been imagined, persistently if often only implicitly, as a sphere of immobility. It is time and history which have claimed 'politics' as their own. As Fabian quotes Ernst Bloch, 'the primacy of space over time is an infallible sign of reactionary language' (Fabian, 1983, p. 37, citing Bloch, 1932/1962, p. 322).

After structuralism

From the point of view of the argument of this book, what *post*-structuralism has most importantly achieved is the dynamisation and dislocation of structuralism's structures. Ironically, temporalisation has opened them up to spatiality – or, at least, it has the potential to do so. It has imbued those structures with temporality and cracked them open to reveal the existence of other voices.

Chantal Mouffe and Ernesto Laclau have been important theorists in this movement. Their aim, in this regard, has been both to open structures up to temporality and to conceive of temporality as open, as involving the potential for the production of the new. The problem of structuralism (and the problem of other forms of temporality too, such as the teleology of certain forms of Marxism) in relation to an opening up to politics is conceived as being causal closure. The aim must therefore be to open up structures through the dislocation which makes politics possible.

Mouffe and Laclau do this in a most productive manner. In its arguments for the openness of temporality, and in its abandonment of the synchrony/diachrony binary, their project of radical democracy is absolutely in tune with the arguments being made here. The crucial recognition, from our point of view, is that the closure of structures is directly related to their a-temporality.

And yet, in spite of all this significant work of reconceptualisation, Laclau, most particularly in his *New reflections on the revolution of our time* (1990), retains a language of space and spatialisation which is unaltered from the earliest structuralism. Temporality is reconceptualised in a liberating manner, but 'space/spatiality' is left relatively unattended. And the terminology of space/spatiality is employed to designate, simply, that which is lacking temporality. It is not reconceptualised in its own right. Structures which are closed (for instance structures of hegemony and of representation) are labelled 'space'. And, correlatively, the notion of spatiality refers above all to a lack of causal openness.

And yet Laclau's approach is both more complex than that and contains within it a contradictoriness which precisely begins to hint at a way out of its own formulation. First, his notion of spatiality refers not to a contemporaneity in a moment of clock/calendar time but to causal closure: that is, not to the instant but to the structuralists' synchrony. Thus certain forms of 'time', those that do not have the characteristic of the production of novelty, are classified by Laclau as space. For instance:

> The representation of time as a cyclical succession, common in peasant communities, is in this sense a reduction of time to space. Any teleological conception of change is therefore also essentially spatialist. (p. 42)

In Laclau's terminology, in other words, what is at issue in the conceptualisation of space is not a lack of 'time' but a lack of 'temporality'. Space is not a-temporal because it presupposes a *coupure* at an instant of clock or calendar time. The crucial characteristic of this definition of space is its causal closure:

> Any repetition that is governed by a structural law of successions is space. (p. 41)

> spatiality means coexistence within a structure that establishes the positive nature of all its terms. (p. 69)

In other words, the causal closure is exactly that of the essential section where 'all the elements of the whole ... are in an immediate relationship with one another' (Althusser, 1970, p. 94). (There is a clear similarity here with Bergson's objection to a notion of temporality which is 'merely a rearrangement of what has been' – Adam, 1990, p. 24.)

However, if this first elaboration by Laclau eventually leads us back to a point we have been at before, his second excursion is more productive. For Laclau (1990) does not use the term 'spatial' only in this way, to refer to a causally closed system. He also, bravely, confronts this usage with what he calls 'physical space'. The relationship turns out to be complex.

To begin with, space and temporality are absolutely opposed:

> dislocation is the very form of temporality. And temporality must be conceived as the exact opposite of space. The 'spatialization' of an event consists of eliminating its temporality. (p. 41)

Then we are assured that this is not a metaphorical use of terminology:

> And note that when we refer to space, we do not do so in a metaphorical sense, out of analogy with physical space. There is no metaphor here. (p. 41)

(At this point we might wonder what kind of space, then, is at issue ...)
Finally, indeed, it is argued that 'physical space' must be temporal too:

> The ultimate failure of all hegemonization, then, means that the real – including physical space – is in the ultimate instance temporal. (p. 42)

This is the kind of resounding QED which begins to gnaw at the foundations of its own demonstration. Its triumphant closure (precisely) reveals the possibility of its deconstruction. On the one hand, certain kinds of time must be classified as space. On the other hand, certain kinds of space (physical space in this instance) must be understood as temporal. In other words, the term 'space' is being mobilised here, not to refer to anything we might understand as being positively spatial (like Laclau's 'physical space'), but rather to designate a lack of (a particular definition of) temporality. What is being referred to is not space as an aspect of space-time, but a-temporal conceptual schema. And Laclau himself implies as much. 'Physical space', too, is temporal. Once again, then, this is space as representation, but from a different angle. This is not the substitution of the trace for the journey but the substitution of the closed coherent system for the inevitable dislocation of the world. Either way, our imaginations of space are seriously diminished.

At one level, then, the problem of Laclau's formulation is 'merely' one of terminology. If he were to drop the equation of the terms space/spatial with causal closure (and hegemonisation–representation), all would be well.

In fact, however, things are not so simple. For the conceptualisation of space in this politically deadening way has reverberations through the rest of the analysis. *First*, 'space' in Laclau's formulation is deprived of any potential for politics. Since it is causally closed it holds open no possibility for genuine change or intervention, for the radically new. 'Politics and space are antinomic terms. Politics only exist insofar as the spatial eludes us' (p. 68). Since, as we have seen, 'space' does not actually refer to *space* this might seem inconsequential as a formulation – except of course that it tends connotatively to perpetuate that view of space *in general* as the realm where nothing happens. *Second*, because space has been characterised in such a derogatory way, the realm of the spatial itself (physical, social space, the space of the human geographers) is rarely directly addressed. And because of this, and *third,* a whole potential field of the sources of dislocation is left unexplored. Since for Laclau 'dislocation is the source of freedom' (p. 60), where freedom is the absence of determination, the necessary unrepresentable 'maladjustment' (p. 42) which provides the possibility for politics, this is not unimportant.

If one wanted to be mischievous one could point to a certain potential circularity:

> insofar as any 'transcendentality' is itself vulnerable, any effort to spatialize time ultimately fails *and space itself becomes an event*. (p. 84; my emphasis)

and again …

> history's ultimate unrepresentability is the condition for the recognition of our radical historicity. It is in our pure condition of event, which is shown at the edges of all representation and in the traces of temporality corrupting all space,

where we find our most essential being, which is our contingency and the intrinsic dignity of our transitory nature. (p. 84)

It is from within this dislocation within the argument of radical democracy itself (or this particular formulation of it) that a thread can be pulled to develop new thoughts. The logic can be pushed beyond its apparent limits. For if space is an event, if traces of temporality corrupt all space, then two things follow: first space becomes as impossible to represent as is temporality (confirming our earlier argument) and second 'space' in the sense that the term was mobilised to indicate a closed and coherent structure, *cannot exist*. Laclau, having defined space as closure, argues that closure is impossible ('the crisis of all spatiality', p. 78). Clearly, one way or another, 'space' must be imagined differently.

The impulse behind Laclau's project is productive and exciting. I would argue that his proposal for a 'radical historicity' could be even more radical were it to be spatialised: that is, were it to recognise from the outset that space is indeed, as he says, 'an event'. But this holding on to a dichotomy between space and time, within which the language of space is reserved for the essentially immobile, is not some idiosyncratic trait. It runs deep through the work of many theorists who have struggled against the stasis of structuralism.

Michel de Certeau is widely cited in the literature on spatiality, particularly urban spatiality. And yet, I would argue, his formulation of the field is hindered by his initial framing device and, moreover, that overarching structure is once again conceptualised, and problematically so, in terms of space and time.

De Certeau's thesis in *The practice of everyday life* (1984) is framed by a contrast between strategies and tactics. A strategy is defined as relating to an already-constructed place, static, given, a structure. Tactics are the practices of daily life which engage with that structure.

This immediately introduces a dichotomy, which might be questioned in its own terms, between structure and agency. It involves a conception of power in society as a monolithic order on the one hand and the tactics of the weak on the other. Not only does this both overestimate the coherence of 'the powerful' and the seamlessness with which 'order' is produced, it also reduces (while trying to do the opposite) the potential power of 'the weak' and obscures the implication of 'the weak' in 'power'. But the issue also runs more deeply, for throughout the book strategies are interpreted in terms of *space* and tactics in terms of *time*:

A strategy assumes a place that can be circumscribed as *proper* (*propre*) … The 'proper' is a victory of space over time. On the contrary, because it does not have a place, a tactic depends on time – it is always on the watch for opportunities that must be seized 'on the wing'. (p. xix; emphasis in the original) strategies pin their hopes on the resistance that the *establishment of a place* offers to the erosion of time; tactics on a clever *utilization of time*, of the opportunities it presents and also of the play that it introduces into the foundations of power. … the two ways of acting can be distinguished according to whether they bet on place or on time. (pp. 38–9; emphasis in the original)

A hundred and one thoughts and objections immediately arise on reading such a passage. It instates a notion of power-relations as simply dichotomised: power versus resistance. Symptomatically, it attempts to escape from an impasse of structuralism (by introducing a notion of resistance) while leaving the structures conceptually intact *and defined as spatial*. And the labelling of this power/resistance binary as spatial/temporal seems to be no more than a resonance from that intellectual history.

Throughout his book de Certeau draws a parallel between the structures of his own analysis and linguistic structures, in particular the distinction between *langue* and *parole*. Indeed, this provocation by the debate over structuralism is explored by Meaghan Morris (1992a) in her 'King Kong and the human fly', which examines de Certeau's account of a visit to the World Trade Center. As I do, she interprets him as struggling to move away from structuralism, and yet …

de Certeau's move from summit to street involves a troubling reinscription of a theory/practice opposition – semantically projected as 'high' versus 'low' ('élite' versus 'popular', 'mastery' versus 'resistance'), 'static' versus 'dynamic' ('structure' versus 'history', 'metanarrative' versus 'story'), 'seeing' versus 'doing' ('control' versus 'creativity', and ultimately, 'power' versus 'know-how') – which actually blocks the possibility of walking away at all. In fact, de Certeau's visit to the World Trade Center is a way of mapping all over again the 'grid' of binary oppositions within which so much of the debate *about* structuralism was conducted (by Sartre and Lévi-Strauss, among others). (p. 13)

Precisely. However, one binary which Morris doesn't mention is that between space and time. De Certeau reinstates that one too. And this is doubly ironic since his whole intention is the opposite. He criticises functionalist organisation, which, 'by privileging progress (i.e. time), causes the condition of its own possibility – space itself – to be forgotten; space thus becomes the blind spot in a scientific and political technology' (1984, p. 95). Here could indeed lie a fault-line in de Certeau's argument which enables it to be levered open and developed.

This is an imagination of power (central bloc versus little tactics of resistance) which maps itself onto the space of the city as similarly divided: the city structure versus the street. Against 'the city as system', the implacable presence of stabilised legibility, is romanticised a mobile 'resistance' of tactics, the everyday, the little people (see, for a particularly clear exposition, de Certeau, pp. 94–8). On the one hand there cannot be such a secure and self-coherent system (the city as synchronic structure), whether we characterise it as space or not. At the very least, even the most monolithic of power-blocs has to be maintained.

On the other hand this central power is understood as removed from 'the everyday' (as opposed to …?), iconically characterised by the street. It is an imagination which has taken a strong hold in urban literature, with its own elaborations of the spatiality of this street as 'the margins', 'the interstitial spaces' and other evocations. At its worst it can resolve into the least politically convincing of situationist capers – getting laddish thrills (one presumes) from rushing about down dark passages, dreaming of labyrinths and so forth. (Is this not itself another form of eroticised colonisation of the city?) As Kristin Ross has asked:

> And what of the street? … The street itself, or at least the backstreets, biways and detours … is the site … of deviance, or (to use the word most popularized by followers of de Certeau), 'resistance'. But resistance to what? In de Certeau movement is escape … (1996, p. 69)

> Criticism derived from de Certeau (that is, much of US 'cultural studies' today) takes capitalism for granted as a kind of forcefield or switchboard that processes meanings; the Salvadoran or Guatemalan selling oranges on the free-ways of Los Angeles becomes a figure of 'resistance' – someone who has appropriated urban space and used it to his own devices, someone thumbing his nose at the 'master planners'. But resistance to what? (p. 71)

What Ross is really worrying about here is the lack of coherence in this resistance ('Tactics add up to no larger strategy', p. 71), and a lack of singular focus (tactics 'are not made to refer back to capital nor to offer any means of understanding the system as a whole', p. 71). This is not my point; it is yet another problematical spatialisation. I am arguing for an abandonment of that dichotomisation between space and time which posits space both as the opposite of time and, equally prob-lematically, as immobility, power, coherence, representation. The significance of this, as the rest of the book will explore, is political.

There is, I think, an irony in the writing of authors such as Laclau and de Certeau (and, as I shall go on to argue, in much of post-structuralism broadly defined). The broad conceptual thrust is to open up the structures of our imag-inations to temporality (Laclau through dislocation, de Certeau through tactics). Yet in the midst of this invigorating concern with time neither author engages in any fundamental critique of the associated terminologies, and con-cepts, of space. In this they are by no means alone. Bergson's *Time and free will* adopts a similar course. Space is a residual category whose definition is derived without much serious thought. Yet one thing which emerges from all this, I would argue, is the interconnectedness of conceptualisations of space and conceptualisations of time. Imagining one in a particular way should, at least 'logically', imply a particular way of thinking about the other. This is not to argue that they are the same, in some easy four-dimensionality. It is to argue that they are integral to each other, which is quite a different proposition.

At minimum, for time to be open, space must be in some sense open too. The non-recognition of the simultaneity of openended multiplicities that is the spatial can vitiate the project of opening up temporality. It cannot be that realm referred to by Foucault: the dead, the fixed; nor can it be the realm of closure, or of static representation. Space is as impossible to represent as is time (though the question of consequence is the representation of time-space). Levering space out of this immobilising chain of connotations both potentially contributes to the dislocations necessary for the existence of the political, and opens space itself to more adequate political address.

4
the horizontalities of deconstruction

The use of the terminology of the spatial to refer to the realm of the immobilised, which was the focus of Chapter 3, does not, however, characterise all post-structuralist writing. There is, of course and most obviously, Foucault's famous reflection: 'Space was treated as the dead, the fixed, the undialectical, the immobile. Time, on the contrary was richness, fecundity, life, dialectic' (1980, p. 70), although the lateness of this retrospection goes some way to confirm that much writing 'after structuralism' retained these conceptual predispositions.

But there is also, more fundamentally, Derrida's recognition of the significance of space/spacing. Unlike Laclau and de Certeau, Derrida does not employ the terminology of space as a simple residual-category negativity of the temporal. He gives it explicit attention in its own right. The very concept of *différance* holds within it an imagination of both the temporal and the spatial (deferral *and* differentiation). Derrida is explicit, too, about certain aspects of space which I would argue are crucial (space as interval, and as holding open the possibility of an open future). Within deconstruction (at least in its theory if not always in its practice), space is explicitly temporalised; changing the 'e' to an 'a' adds time to space. 'Dissemination' 'marks an irreducible and *generative* multiplicity' (1972/1987, p. 45; emphasis in the original), only *différance* is fully historical. This mobilisation and fracturing of structures both questions pretensions to integrity and self-presencing and overcomes the impasse of *langue* versus *parole*. For Derrida spacing is fundamental to difference/*différance*. It enables the opening up of the usual meaning of 'history'. In *Of grammatology* he writes, 'The word "history" doubtless has always been associated with the linear consecution of presence' (cited in 1972/1987, p. 56). One might query the all-too-easy mobilisation of 'always', but the sentiment is well taken. And this linearity of the (then) hegemonic meaning of history is argued to have a whole set of further implications ('an entire *system* of implications' – 1972/1987, p. 57; emphasis in the original), including teleology, continuity and the assumption of an interiorised accumulation of meaning. All this is entirely in the spirit of what I have been trying to work towards here. Indeed, Marcus Doel (1999) has argued that post-structuralism is already spatial. It is, he argues, precisely the event of space, of spacing, which deconstructs all hypothesised integrities.[13] My argument is rather that post-structuralism *could* very easily be spatial (in the way I mean that term here). But, as Derrida himself points out, for deconstruction to live, and particularly when it is being transported into new areas, it will need to be transformed. Just as in the engagements with Bergson, structuralism and Laclau, the sympathetic trick is to work within but to emerge, maybe, with something appropriately different.

Deconstruction has throughout been strongly concerned with textuality; with speech and writing, and with texts. These were the debates within which it established its own differentiation. As a mode of working it has subsequently been argued to extend more widely (though, as Derrida says, it is with 'words' that he himself feels most at home). There has been, none the less, a shift from a focus on what came to be called texts 'in the narrow classic sense' towards an expansion of scope in later works. As Derrida puts it at one point, 'even if there is no discourse, the effect of spacing already implies a textualization' (1994, p. 15). Representation again, in a sense, but the aim here is to challenge the pretensions to closure of the text.

Thus, as the argument, and the language within which deconstruction has pursued its case, have evolved there has been a claim for increasing generalisability. The proposition which emerges is that 'the world is like a text'. Here, instead of representation being imagined as spatialisation – 'spacing … implies … textualisation' – the movement is reversed. As with every proposition, this is a statement with a history, its own process of differentiation. For those of us who did not follow that particular historical trajectory (whose engagements and differentiations have been otherwise) an equivalent (but not identical) proposition might be that texts are really just like the rest of the world. But of course the trajectory of engagement, the sequence of repetition and differentiation, has effects. The direction from which you come at an argument influences its form. 'The world is like a text' is a proposition quite distinct from 'texts are just like the rest of the world'. There are real reasons for being attentive to the routes of thought's imagination.

There is, for instance, a residual but persistent 'horizontality' about the approach of deconstruction which makes it difficult for it to handle (or, rather, to provoke an imagination of) a spatiality which is fully integral within space-time. Texts present themselves as two-dimensional structures; horizontal coherences/integrities which can be shown, through deconstruction, not to be coherent at all. There is no doubt about the liberatory aspects of this manoeuvre. And indeed what I am trying to argue here in relation to space shares much of the same impetus. The deconstruction of presumed horizontal integrities chimes well with the critique of place as internally coherent and bounded (Massey, 1991a). The emphasis on horizontality can be interpreted as (and in some senses and circumstances it actually *is*) a turn towards spatiality and a spatiality, what's more, which is open and differentiated. It seems, therefore, ironic – if not downright churlish – to raise any objection. Yet perhaps there is in this formulation (this mind's eye imagination of the intellectual task at hand) *too much* emphasis on the purely horizontal and too little recognition of the multiple trajectories of which that 'horizontality' is the momentary, passing, result. As John Rajchman (1998) observes, in a related querying of the constructiveness of the horizontal view, collage and superposition, once celebrated, have become obstacles (p. 9; see also his essay *Grounds* in the same volume). The nature of (the practice of) deconstruction leads it to emphasise the aspect of *différance* which is differentiation over that which is differral.

This is not inherent in the conceptual structure of deconstruction. Derrida frequently stresses the joint productivity of spatial and temporal dimensions. The long interview with Jean-Louis Houdebine and Guy Scarpetta (Derrida, 1972/1987, pp. 37–96) exemplifies the complex of issues at stake. In a note to this discussion (footnote 42, pp. 106–7) he writes: 'spacing is a concept which also, but not exclusively, carries the meaning of a productive, positive, generative force. Like *dissemination*, like *différance* it carries along with it a *genetic* motif: it is not only the interval, the space constituted between two things (which is the usual sense of spacing), but also spa*cing*, the operation. ... This movement is inseparable from temporization – temporalization (see *"La différance"*) and from *différance*' (emphases in the original). Spacing is here *both* (what we would normally term) spatial *and* temporal.

And yet, the way in which Derrida conceives of this processual/temporal aspect of spacing poses problems in its turn. The ellipsis in the above quotation, when filled in, provides a hint. Here, 'the operation' (the process which is spacing) is defined as 'the movement of setting aside' (p. 106) and the passage continues: 'It [the movement of spacing] marks *what is set aside from itself*, what interrupts every self-identity, every punctual assemblage of the self, every self-homogeneity, self-interiority' (p. 107; my emphasis). Now, there are two things going on here, two forms of what might be called negativity, both of which are problematical for an analysis of social, physical, space.

The first was just highlighted in italics: the conceptualisation of spacing as an act of (attempted) setting-aside, the process of expulsion supposedly necessitated by the aim of constructing a self-identity (here defined in terms of homogeneity, self-interiority, etc). The focus is on rupture, dislocation, fragmentation and the co-constitution of identity/difference. Conceptualising things in this manner produces a relation to those who are other which is in fact endlessly the same. It is a relation of negativity, of distinguishing *from*. It conceives of heterogeneity in relation to internal disruption and incoherence rather than as a positive multiplicity. It is an imagination from the inside in. It reduces the potential for an appreciation of a positive multiplicity beyond the constant reproduction of the binary Same/Other. This is both politically disabling and problematical for a rethinking of the spatial. Politically, as Robinson (1999) argues, in some of this

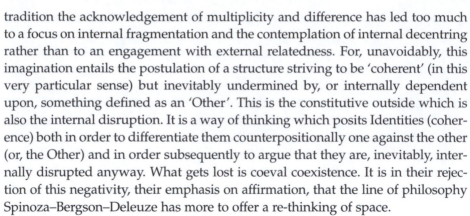

tradition the acknowledgement of multiplicity and difference has led too much to a focus on internal fragmentation and the contemplation of internal decentring rather than to an engagement with external relatedness. For, unavoidably, this imagination entails the postulation of a structure striving to be 'coherent' (in this very particular sense) but inevitably undermined by, or internally dependent upon, something defined as an 'Other'. This is the constitutive outside which is also the internal disruption. It is a way of thinking which posits Identities (coherence) both in order to differentiate them counterpositionally one against the other (or, the Other) and in order subsequently to argue that they are, inevitably, internally disrupted anyway. What gets lost is coeval coexistence. It is in their rejection of this negativity, their emphasis on affirmation, that the line of philosophy Spinoza–Bergson–Deleuze has more to offer a re-thinking of space.

There is a hilarious engagement in Derrida's interview with Houdebine and Scarpetta which revolves around this distinction between negative difference and positive heterogeneity. For Derrida spac*ing* is integral to the constitution of difference. Towards the end of his conversation with Derrida, Houdebine tries to specify this a little further (Derrida, 1972/1987, p. 80 *et seq*). Derrida doesn't grasp the point of the question, and Houdebine tries again: 'No, that is not what I said: let me rephrase the question: is the motif of heterogeneity entirely covered by the notion of spacing? Do not *alterity* and *spacing* present us with two moments not identical to each other?' (p. 81; emphasis in the original). The two men continue to talk past each other in the interview itself and then again in the footnotes, which contain reflections on the interview (see pp. 106–7) and in a subsequent exchange of letters (pp. 91–6). In his letter, Houdebine insists again that

> everything derives from my question on the motif of *heterogeneity*, a motif that I think is irreducible to the single motif of spacing. That is, the motif of heterogeneity indeed implies, in my opinion, the two moments of *spacing* and of *alterity*, moments that are in effect indissociable [here he is saying 'yes yes' to Derrida who had earlier insisted on this point which is not the point], but that are also not to be identified with each other. (p. 91; emphasis in the original)

In the midst of all the confusion, there is then a hint of what may be a source of Derrida's continuing to read the issue differently from Houdebine. It comes at a point where Houdebine refers back to something Derrida had said earlier: 'spacing', he had said,

> is the index of an irreducible exterior, and at the same time of a *movement*, a displacement that indicates an irreducible alterity. I do not see how one could dissociate the two concepts of spacing and alterity. (p. 81, emphasis in the original)

This, for me, precisely indicates a problem. Difference and multiplicity are here intimately associated through a process, and that process is one of displacement and exteriorisation (elsewhere abjection, repression, etc.). The coexistence of others, and the specification of their 'difference', are recognised through the one process of their being 'set aside' (p. 107). It is an imagination which, in spite of itself, starts from the 'One' and which constructs negatively both plurality and difference. A touch of exasperation seems to infiltrate Houdebine's letter:

> it remains [the case] that the motif of heterogeneity is not reduced to, is not exhausted by this 'index of an irreducible exterior.' It is *also the position of this alterity as such*, that is, the position of a 'something' (a 'nothing') [i.e. 'spacing designates *nothing*, ... but is the index of an irreducible exterior', p. 81] that *is not nothing*. (p. 92; all emphases in the original, text in square brackets added)

Quite so. And Houdebine insists: 'The complete development of the motif of heterogeneity thus obliges us to go to the positivity of this "nothing" designated by spacing' (p. 92).[14] By page 94 an accommodation is being arrived at. Says Derrida:

> The irreducibility of the other is marked in spacing in relation to what you seem to designate by the notion of 'position' [a 'something' ... (the *position* of an irreducible alterity) (Houdebine, on p. 92; emphasis in the original)]: in relation to our discussion of the other day, *this is the newest and most important point, it seems to me* ... (p. 94; my emphases)

The lineaments of this delightful philosophical engagement contain much that is relevant to an alternative imagination of space. The significance of a recognition of the *fact* of spacing. The integration within this of both space and time. The wrestling over how the process of difference/heterogeneity is to be conceptualised. The contrast between the negativity (expulsion, abjection ...) of Derrida's view and Houdebine's search for 'positivity'. Even, perhaps, the very difficulty of the argument. Derrida indeed acknowledges its significance. It was in recognition of this significance that he closed the communication with the proposal that it be entitled *Positions*. And it was.

Position, location, is the minimal order of differentiation of elements in the multiplicity that is co-formed with space.

But there is a second aspect of negativity: the constant use of a language of disruption, dislocation, decomposition and so forth. Derrida has, of course, endlessly addressed aspects of this accusation. He has argued, rightly, that this was precisely the task which had initially to be accomplished. 'Structures were to be undone, decomposed, desedimented' (Kamuf, 1991, p. 272, where Derrida is precisely reflecting on the historical placing of his work). In the terms of the earlier discussion, it was a question of undoing closure. He has also argued that it is 'not a question of junking concepts, nor do we have the means to do so', and in 'The case of the concept of *structure* ... Everything depends on how one sets it to work' (1972/1987, p. 24; emphasis in the original). The way forward is to transform concepts and, little by little, to produce new configurations: this is '*la double séance*', a writing that is both within and striving to escape the inherited infrastructure of the imagination. An attempt simply to make a break for it will often (Derrida, typically, says 'always') lead to the reinscription of the supposedly new ideas within the same old cloth (p. 24). The aim must be 'to transform concepts, to displace them, to turn them against their presuppositions, to reinscribe them in other chains, and little by little to modify the terrain of our work and thereby produce new configurations' (p. 24). In the end, we might, as Derrida writes quite wonderfully, indulge 'the desire to escape the combinatory itself, to invent incalculable choreographies' (1995, p. 108; cited in Doel, 1999, p. 149). But this is precisely the difficulty: that that process of invention seems itself to be constrained by deconstruction's horizontality and negativity, by its embeddedness in an intellectual trajectory which has emerged from a concern with the textual (and in some guises the psychoanalytic). It is harder to get from deconstruction to that understanding of the world as becoming, as the positive creation of the new, which is so central to the philosophies of Spinoza–Bergson–Deleuze. It is also, therefore, unable to generate a recognition of space as the sphere of coexisting multiplicity, space as a simultaneity of stories-so-far. On its own the viewpoint of deconstruction is not enough to achieve that necessary transcribing of space from the chain stasis/representation/closure into an association with openness/unrepresentability/external multiplicity. What is at issue is almost like a shift of physical position, from an imagination of a textuality *at which one looks*, towards recognising one's place *within* continuous and multiple processes of emergence.

And perhaps one thing which makes this a particularly tricky manoeuvre for deconstruction in relation to a reconceptualisation of spatiality is that other inheritance: of the association between text/writing and space. To shift the imagination from a mission to disrupt the supposed integrity of spatial structures towards an ever-moving generative spatio-temporal choreography is peculiarly difficult where the very notion of the dislocation of structures has so frequently been itself translated as the dislocation *of* space *by* time. As Derrida himself writes (see above), 'the effect of spacing already implies a textualization' (1994, p. 15). Coming at it from another angle hints at what it might mean to argue *not* that the world (space-time) is like a text but that a text (even in the broadest sense of that term) is just like the rest of the world. And so might be avoided the longstanding tendency to tame the spatial into the textual.

5
the life in space

Almost all the lines of thought explored in Part *Two* have encompassed more than one understanding of space. In excavating this the aim has been both to point to the problematic repercussions of some associations and to emphasise the potential of alternative views. The hope is to contribute to a process of liberating space from its old chain of meaning and to associate it with a different one in which it might have, in particular, more political potential.

The argument started from the position that space is a discrete multiplicity, but one in which the elements of that multiplicity are themselves imbued with temporality. A static contemporaneity was rejected in favour of a dynamic simultaneity. Another form of closing down an appreciation of the dynamic multiplicity that is space was argued to be its imagination as an immobile closed system. The argument here is instead to understand space as an open ongoing production. As well as injecting temporality into the spatial this also reinvigorates its aspect of discrete multiplicity; for while the closed system is the foundation for the singular universal, opening that up makes room for a genuine multiplicity of trajectories, and thus potentially of voices. It also posits a positive discrete multiplicity against an imagination of space as the product of negative spacing, through the abjection of the other. It rejects, also, Laclau's use of 'space' to refer to static closure ('the cemetery or the lunatic asylum', Laclau, 1990, p. 67) in favour of his recognition that space itself is an event.

On this reading neither time nor space is reducible to the other; they are distinct. They are, however, co-implicated. On the side of space, there is the integral temporality of a dynamic simultaneity. On the side of time, there is the necessary production of change through practices of interrelation. 'The connections among things alone make time' (Latour, 1993, p. 77 – although one might wish also to recognise the co-production of the entities in the connections); 'Time is … a provisional result of the connection among entities' (p. 74). Change requires interaction. Interaction, including of internal multiplicities, is essential to the generation of temporality (Adam, 1990). Indeed, were we to assume the unfolding of an essentialist identity the terms of change would be already given in the initial conditions. The future would not in that sense be open. And for there to be interaction there must be discrete multiplicity; and for there to be (such a form of) multiplicity there must be space. Or, as Watson (1998) in his exploration of 'the new Bergsonism' writes, that tradition understands autopoiesis in terms of structural coupling between dissipative structures. Deleuze's 'radical empiricist' conjuncturally determined play between internal and external relations catches at this (Hayden, 1998). We cannot 'become', in other words, without others.[15] And it is space that provides the necessary condition for that possibility. Bergson, in response to his own question 'what is the role of time?', replied 'time prevents everything from being given at once' (1959, p. 1331). In this context the 'role of space' might be characterised as providing the condition for the existence of those relations which generate time.

This must, however, be distinguished from the claim that 'space is important because it contributes to the temporally new'. This is the case, and the argument will be put in what is to come. But the position here goes further than that. Indeed, Grossberg (1996) has written ironically of some of the ways in which attempts have been made to rescue space from a perceived deprioritisation, and 'The first [of these] puts space to work in the service of time; that is, it makes the power of space instrumental, raising important questions of how power uses, organises and works through space, yet reducing it to its role in securing the demands of temporal power (i.e. the reproduction of structure)' (p. 177). The argument here concerns the *mutual* necessity of space and time. It is on both of them, necessarily together, that rests the liveliness of the world.

These arguments are by no means all new. I have precisely been trying to draw on the sometimes underplayed insights of others. Moreover, when stated like this the response might be, 'of course; this is obvious'. Yet in many current discourses space is practised and imagined quite otherwise. In particular, quite different imaginaries and engagements of space are mobilised as foundations within political questions. Part *One* has already hinted at this and it will be taken up directly in what is to come. The aim here has been to prepare some of the ground.

Moreover, this issue of how we might imagine space intersects with the question of subjectivity itself. Elizabeth Grosz, in *Space, time, and perversion*, links into a number of the arguments here when she writes:

> Newtonian mechanics, like Euclidean geometry, reduces temporal relations to spatial form insofar as the temporal relations between events are represented by the relations between points on a straight line. Even today the equation of temporal relations with the continuum of numbers assumes that time is isomorphic with space, and that space and time exist as a continuum, a unified totality. *Time is capable of representation only through its subordination to space and to spatial models.* (1995, p. 95; my emphasis)

As has been seen, the most common argument against this procedure is driven by the damage that it does to time: that it turns it into a discrete multiplicity.

My argument has been that it also does damage to space, insofar as that discrete multiplicity is imagined also as static. Grosz, however, develops another line of argument, which relates to imaginations of subjectivity. She writes 'there is an historical correlation between the ways in which space (and to a lesser extent, time) is represented, and the ways in which subjectivity represents itself' (p. 97). Then, through the work of Irigaray (1993), she posits a connection to interiority and exteriority, where space is conceived as the mode of exteriority and time as the mode of interiority. This is a persistent philosophical theme. While Irigaray draws on ancient theology and mythology: 'In Kant's conception, too, while space and time are *a priori* categories we impose on the world, space is the mode of apprehension of exterior objects, and time a mode of apprehension of the subject's own interior' (p. 98).

Grosz then connects this time–space distinction to the constitution of gender:

> This may explain why Irigaray claims that in the West time is conceived as masculine (proper to a subject, a being with an interior) and space is associated with femininity (femininity being a form of externality to men). Woman is/ provides space for man, but occupies none herself. Time is the projection of his interior, and is conceptual, introspective. The interiority of time links with the exteriority of space only through the position of God (or his surrogate, Man) as the point of their mediation and axis of their coordination. (1995, pp. 98–9)

Gillian Rose (1993), again drawing on Irigaray, has also analysed these gendered distinctions between space and time, and there are significant connections to the argument being made here. It has already been seen, for instance, how Prigogine and Stengers point to some philosophers' interiorisation of time as irreversible in the face of natural science's insistence on its 'objective' reversibility. Bergson started from experience; it was experience which challenged the proposed divisibility of time; experience was duration. And the insistence on analysing time in this way has been a continuous thread (see, as a recent example, Osborne, 1995). Even philosophers who are aware of embodiment as an element in an interconnected (that is, spatial) world can none the less stress this purely temporal aspect of subjectivity. Thus, from a different trajectory again, Merleau-Ponty writes 'we must understand time as the subject and the subject as time' (1962, p. 422, cited in Mazis, 1999, p. 231), or again 'the perceptual synthesis is a temporal synthesis, and subjectivity at the level of perception is nothing but temporality' (p. 332, cited in Mazis, p. 234). 'The smallest possible experience is therefore a difference or moment in the experienced *passage of time*,' writes Deleuze (1953/1991, pp. 91–2; my emphasis); 'not all ideas give the quality of spatial extension, but all atoms [of experience] give the quality of time in which they occur' (Goodchild, 1996, p. 17). 'So', comments Goodchild, 'Deleuze's empiricism is tied not to a naive atomistic conception of matter or experience, but to *time* as the basis of both meaning and experience' (1996, p. 17; my emphasis). Grossberg, indeed, has made the significant claim that 'The bifurcation of time and space, and the privileging of time over space, was perhaps the crucial founding moment of modern philosophy [in a footnote he makes clear that it is 'the separation' of time and space that is the crucial issue]. It enabled the deferral of ontology and the reduction of the real to consciousness, experience, meaning and history' (1996, p. 178). Moreover this assumption of the pure temporality of interiority is in turn connected to the counterpositioning of space not just as external but as *material*. As Boundas comments, in relation to Bergson–Deleuze's distinction between discrete and continuous: 'In a sense, the great dualism inherited from the classical rationalists and empiricists – matter and mind – is repositioned now on the distinction between duration and space' (1996, p. 92).

There are two things going on here. First the analysis of the temporal as interior. And second the understanding of interiority as purely temporal. The latter is, as Grosz puts it, one of 'the ways in which subjectivity represents itself' and that in turn, as she argues, has been correlated with the ways in which space is understood.

Maybe then if we think and practise space differently it will reverberate in these other realms too. One line of critique has revolved around a kind of philosophical miserablism which has on occasions characterised the preoccupation with time. In stark contrast to the evocations of Bergson–Deleuze, it has been argued that much writing on time, and its frequent association with interiority, derives from an obsessive fear of death (see, for instance, Cavarero, 1995). There is also that line of questioning, especially from feminist philosophers, which puts the political arguments for understanding identity/subjectivity in a more strongly relational manner. It harks back indeed to the relational construction of space. Thus Moira Gatens and Genevieve Lloyd (1999) have drawn on Spinoza to explore the relational construction of subjectivity, the inseparability of individuality and sociability. This releases our imaginations. For if experience is not an internalised succession of sensations (pure temporality) but a multiplicity of things and relations, then its *spatiality* is as significant as its temporal dimension. This is to argue for a way of being and thinking otherwise – for the imagination of a more open attitude of being; for the (potential) outwardlookingness of practised subjectivity. Thus as Bergson's thought evolved, 'Duration seemed to him to be less and less reducible to a psychological experience and became instead the variable essence of things, providing the theme of a complex ontology. But, simultaneously, space seemed to him to be less and less reducible to a fiction separating us from this psychological reality, rather, it was itself grounded in being' (Deleuze, 1988, p. 34). The two evolutions are related. As Deleuze cites it: 'Movement is no less outside me than in me; and the Self itself in turn is only one case among others in duration' (p. 75). As Lloyd argues 'For [Spinoza], we do not gain our true selves by withdrawing behind our frontiers. We become most ourselves by opening out to the rest of nature ... these two dimensions of selfhood: the self's relations to the spatial world, in the here and now, and its relations to time. His dynamic physics of bodies provides the nexus between the two … an internal multiplicity of self-hood' (1996, pp. 95–7). Bergson wrote of imaginative leaps: in relation to memory, of placing ourselves 'at once' in the past; in relation to language, of jumping into the element of sense. Is the same leap possible into spatiality? Can one 'throw oneself into spatiality?' (Grosz, 2001, p. 259). Not only, then, duration in external things but also a spatialisation of being in response.

Conceiving of space as a static slice through time, as representation, as a closed system and so forth are all ways of taming it. They enable us to ignore its real import: the coeval multiplicity of other trajectories and the necessary outward-lookingness of a spatialised subjectivity. In so much philosophy it is time which has been a source of excitement (in its life) or terror (in its passing). I want to argue (and setting aside for the moment that we should not separate them like this) that space is equally exhilarating and threatening.

If time is to be open to a future of the new then space cannot be equated with the closures and horizontalities of representation. More generally, if time is to be open then space must be open too. Conceptualising space as open, multiple and relational, unfinished and always becoming, is a prerequisite for history to be open and thus a prerequisite, too, for the possibility of politics.

In a fascinating article, Lechte (1995) also associates 'science' with 'writing' and both of them in turn with space. His argument is that – now – both science (as a result of the new discourses of chance, chaos etc.) and writing (as a result of post-structuralism and deconstruction) have inevitable elements of indeterminacy. He concludes: 'If postmodern science takes us to the limits of knowledge and the beginning of chance, if it discovers that non-knowledge (as the undecidable, as uncertainty, as indeterminacy) is structurally inescapable, what it also discovers … is that through space, writing is tied to science; for writing is also indeterminate' (p. 110). My own reservations about the nature of this reliance on science will be explored in Chapter 11. Nevertheless, I do agree with Lechte's very final sentence: 'The political implications of this are perhaps still to be recognized' (p. 110).

Part *Three*
Living in spatial times?

Part *Two* reflected upon some of the ways in which, through philosophical debates, 'space' has come to have attached to it a range of unhelpful associations which hinder a full recognition of the challenge posed by practical socio-political space. More positively, what emerged was an argument for space as the dimension of a dynamic simultaneous multiplicity. It is with some current, and significant, imaginaries of that socio-political space that this Part now engages, with a particular focus on imaginations of the current era as supposedly 'spatial' and 'globalised'. Underlying these accounts, again, are conceptualisations of space which need to be questioned. For once again they are means of avoiding the real challenge thrown up by the spatial; indeed they are covert means of legitimating its suppression.

Part *Two* wrestled with space as a simultaneity of multiple trajectories. Recognition of that should in principle establish space as posing the question, the challenge, of contemporaneous processual existence. However, in different ways many of the hegemonic discourses and practices explored here avoid that challenge: by convening spatial multiplicity into temporal sequence; by understanding the spatial as a depthless instantaneity; by imagining 'the global' as somehow always 'up there', 'out there', certainly somewhere else. Each is a means of taming the spatial. What all of these spatial (I would call them anti-spatial) strategies do is evade that challenge of space as a multiplicity. And this raises the aspect of practised space which is its relational construction; its production through practices of material engagement. If time unfolds as change then space unfolds as interaction. In that sense space is the *social* dimension. Not in the sense of exclusively human sociability, but in the sense of engagement within a multiplicity. It is the sphere of the continuous production and reconfiguration of heterogeneity in all its forms – diversity, subordination, conflicting interests. As the argument develops, what begins to be addressed is what that must call forth: a relational politics for a relational space.[1]

6

spatialising the history of modernity

If once it was 'time' that framed the privileged angle of vision, today, so it is often said, that role has been taken over by space. The responses have ranged from revelry to fear. One of the moving forces in social science thinking in recent years has been an urge to respond positively: to 'spatialise'. For reasons which range from a deeply political desire to challenge old formulations, through a characterisation of 'postmodern' times as 'spatial rather than temporal', to a surprisingly insouciant, and recent, recognition of the geographical nature of society, much serious attention has been devoted to what has been called 'the spatialisation of social theory'.

One productive example of this has been the postcolonial concern to rework the sociological debates over the nature of modernity and its relation to globalisation. Indeed, for a number of authors 'globalisation' has been the prime form taken by this effort to spatialise sociological thinking. The collection by Featherstone, Lash and Robertson (1994) both makes this point and contains good examples of such spatialisation in practice. Telling a story of globalisation has been used to spatialise the story of modernity. Moreover – and this is the important point – this spatialisation has had *effects* on the concept of modernity and severely dislocated the previous story of its unfolding. Stuart Hall indeed argues that this is one of the main contributions of the postcolonial critique:

> It is the retrospective re-phrasing of Modernity within the framework of 'globalisation' … which is the really distinctive element in a 'postcolonial' periodisation. In this way, the 'post-colonial' marks a critical interruption into that whole grand historiographical narrative which, in liberal historiography and Weberian historical sociology, as much as in the dominant traditions of western Marxism, gave this global dimension a subordinate presence in a story which could essentially be told from within its European parameters. (1996, p. 250)

The implications of spatialising/globalising the story of modernity are profound. The most obvious effect, which has indeed been the main intent, is to rework modernity away from being the unfolding, internal story of Europe alone. The aim has been precisely to decentre Europe. Thus: 'This re-narrativisation displaces the "story" of capitalist modernity from its European centering to its dispersed global "peripheries"' (p. 250). 'Colonisation' becomes more than a kind of secondary by-product of events in Europe. Rather 'it assumes the place and significance of a major, extended and ruptural world-historical event' (p. 249). There is the possibility here, moreover, of a further reformulation. Not only should the European trajectory be 'decentred' it could also be recognised as merely one (though most certainly in military and other terms the most powerful) of the histories being made at that time. This is the multiplicity which is the burden of Eric Wolf's magnificent book *Europe and the people without history* (1982). It is the *meeting-up* of Moctezuma and Cortés. It implies (it could imply) a different view of space itself. It is a move away from that imagination of space as a continuous surface that the coloniser, as the only active agent, crosses to find the to-be-colonised simply 'there'. This would be space not as smooth surface but as the sphere of coexistence of a multiplicity of trajectories.

Moreover, once the multiplicity of trajectories has been recognised, a further effect of spatialising in this way the story of modernity becomes clear. Once understood as more than the history of Europe's own adventures, it is possible to appreciate how the previous way of telling the story (with Europe at its centre) was powered by the way in which the process was experienced *within* Europe; told through the experience of exploration outward from Europe; told from the point of view of Europe as the protagonist. Spatialising that story enables an understanding of its positionality, its geographical embeddedness; an understanding of the spatiality of the production of knowledge itself.

Further, retelling the story of modernity through spatialisation/globalisation exposed modernity's preconditions in and effects of violence, racism and oppression. It is here that the oft-told story of the question posed to modernity by Toussaint l'Ouverture is relevant (Bhabha, 1994). Toussaint l'Ouverture, leader of rebel slaves, had the principles of the French Revolution (modernity) always in his mind. C.L.R. James writes: 'What revolutionary France signified was perpetually on his lips, in public statements, in his correspondence ... If he was convinced that San Domingo would decay without the benefits of the French connection, he was equally certain that slavery could never be restored' (1938, p. 290). He was, of course, 'wrong'. As Bhabha puts it, he had to grasp 'the tragic lesson that the moral, *modern* disposition of mankind, enshrined in the sign of the Revolution, only fuels the archaic racial factor in the society of slavery', and Bhabha asks 'what do we learn from that split consciousness, that "colonial" disjunction of modern times and colonial and slave histories ...?' (1994, p. 244). In other words, the (some of the) material preconditions and effects of the project of modernity, when brought to light by this spatial opening-out, undermine the very story which it tells about itself: 'This re-narrativisation displaces the "story" of capitalist modernity from its European centering to its dispersed global "peripheries"; from peaceful evolution to imposed violence' (Hall, 1996, p. 250). The exposure of those preconditions and effects revealed modernity as precisely being also about the establishment of an enunciative position which (i) although particular, made a claim for universality, but which (ii) was not to be (could not be) in fact universalised or generalised. More complexly, modernity, here in the shape of the French Revolution, opened up the possibility of Toussaint l'Ouverture's question; and the Haitian slave rebellion thus multiplies beyond Europe the trajectories through which modernity was made. In other words, one of the effects of modernity was the establishment of a particular power/knowledge relation which was mirrored in a geography that was also a geography of power (the colonial powers/the colonised spaces) – a power-geometry of intersecting trajectories. And in the postcolonial moment it is that which has come home to roost. For exposing that geography – by the raising of voices located outside of (although geographically often within) the accepted speaking-space of modernity, by insisting on the multiplicity of trajectories – has helped also to expose and undermine the power/knowledge relation.

In all these ways, then, the globalisation/spatialisation of the story of modernity has provided a commentary upon, and thereby challenged, *both* a system of rule *and* a system of knowledge and representation. And both the system of rule and the system of power/knowledge had very definite geographies. Spatialising the story of modernity (both in revealing its operational spatialities and in opening it up to enable the presence of a multiplicity of trajectories) has had effects – it has not left the story the same.

Moreover, within the history of modernity there was also developed a particular hegemonic understanding of the nature of space itself, and of the relation between space and society.[2] One characteristic of this was an assumption of isomorphism between space/place on the one hand and society/culture on the other.[3] Local communities had their localities, cultures had their regions and, of course, nations had their nation-states. The assumption was firmly established that space and society mapped on to each other and that together they were, in some sense 'from the beginning', divided up. 'Cultures', 'societies' and 'nations' were all imagined as having an integral relation to bounded spaces, internally coherent and differentiated from each other by separation. 'Places' came to be seen as bounded, with their own internally generated authenticities, and defined by their difference from other places which lay outside, beyond their borders. It was a way of imagining space – a geographical imagination – integral to what was to become a project for organising global space. It was through that imagination of space as (necessarily, by its very nature)

divided/regionalised that the (in fact particular and highly political) project of the generalisation across the globe of the nation-state form could be legitimated as progress, as 'natural'. And it continues to reverberate today. Even where there is discussion (and where isn't there these days?) of the opening of borders, of the 'new' space of flows, of the transgressing of every boundary in sight … there is still often alongside it an assumption that once (once upon a time) those boundaries were *im*permeable, that there was no transgression. This is an attitude, a cosmology, reflected in all those nostalgic responses to globalisation which mourn the loss of the old spatial coherences. It is a nostalgia for something that did not exist (see also Low, 1997; Weiss, 1998).[4] It is an imagination which, having once been used to legitimate the territorialisation of society/space, now is deployed in the legitimation of a response to their undoing; a response to 'globalisation' (a term which will be examined later but to be read here in its simple sense of increasing global contacts and flows) which consists in retreating into its supposed opposite: nationalisms and parochialisms and localisms of all sorts. This response is not 'backward-looking' (the charge most frequently levelled); it is looking backwards to a past that never was.

It is a response which takes on trust a story about space which in its period of hegemony not only legitimised a whole imperialist era of territorialisation but which also, in a much deeper sense, was a way of taming the spatial. This is a representation *of* space, a particular form of ordering and organising space which refused (refuses) to acknowledge its multiplicities, its fractures and its dynamism. It is a stabilisation of the inherent instabilities and creativities of space; a way of coming to terms with the great 'out there'. It is this concept of space which provides the basis for the supposed coherence, stability and authenticity to which there is such frequent appeal in discourses of parochialism and nationalism. It is this understanding of space which was at work in the third rumination (of 1989 and all that) in the opening chapter. And it provides, too, the basis for much more ordinary notions – persistent and everyday – that 'place', or locality (or even 'home') provides a safe haven to which one can retreat. What was evolved within the project of modernity, in other words, was the establishment and (attempted) universalisation of a way of imagining space (and the space/society relation) which underpinned the material enforcement of certain ways of organising space and the relationship between society and space. And it is still with us today.

It was, moreover, a conceptualisation of space largely endorsed by the social sciences. As Gupta and Ferguson (1992) argue: 'Representations of space in the social sciences are remarkably dependent on images of break, rupture, and disjunction … The premise of discontinuity forms the starting point from which to theorize contact, conflict, and contradiction' (p. 6).

The starting point, in other words, was (is still) very often an imagination of space as already divided-up, of places which are already separated and bounded. Walker (1993) has argued a similar position in relation to the nation-state, and the formulation of the notion of 'place' and of the relation of place to culture and society has had a similar career. Giddens, among others, has pronounced upon the changing relation between 'space' and 'place'. In 'premodern' societies, Giddens (1990) asserts, space was as local as place. Then, with modernity, came the separation of the two: space as the outside of a place which was 'specific, concrete, known, familiar, bounded' (Hall, 1992, characterising Giddens). Today that relationship between space and place, says Giddens, is breaking down, and he is widely cited in the matter.

Now, a lot depends here on how this argument is read. If Giddens is rehearsing the dominant *discourse* of space and place under modernity (and in the West, we should add), then he has certainly captured a common understanding. But that discourse can itself be questioned. Most importantly, it makes assumptions about 'premodern' societies and their relation to space which have been put under serious challenge. Oakes (1993), in his research on place identity in China, precisely questions the supposed past unity of space and place, and the currently much talked-of contrast between a past 'space of places' and a supposedly new 'space of flows': 'In claiming that "the old identity between people and places" has disappeared, there is surprisingly little historical analysis … when was the old community ever "spatially circumscribed"?' (p. 55). And he argues from his own work in China that in the past 'Distinctive cultural spaces were maintained … through connections rather than disjunctions … "locality" is simply a contingent component of that "space of flows" rather than its antithesis' (p. 63).

There are a number of distinct points here. *First* that the evidence for past cultural isolates, and any simple conjunction of space and place, is under challenge. And under challenge too, therefore, is the kind of neat periodisation schematised by Giddens and others (which is by no means to say that there have not been changes). *Second*, that that way of thinking in terms of space-divided-up is a product of modernity's own project (and a source of some of its subsequent anxieties). And *third* that the source of cultural specificity does not lie only in spatial isolation and the emergent effects of 'internal' processes of articulation (where the definition of 'internal' may vary) but importantly also in interactions with the beyond. It is such internal articulation which domesticates (sometimes) the products of interaction, which enables even quite recent cultural imports so easily to be absorbed into the quintessential characteristics of authenticity (the English cup of tea, the Italian pasta which arrived in Italy from China, and so forth).

The anthropological work of Gupta and Ferguson pursues these arguments and links them to notions of identity. Central to their project is the need to challenge the assumed isomorphism of space, place and culture. On the one hand that means abandoning 'the premise of discontinuity' (that is, taking as one's starting point an imagination of space as divided up) and on the other hand it means 're-thinking difference through connection' (Gupta and Ferguson, 1992, p. 8).

Using the example of how 'the Bushmen' came to be Bushmen (through a never-isolated, never-unchanging process of the production of cultural differentiation in interrelational space), they argue that 'Instead of assuming the autonomy of the primeval community, we need to examine how it was formed *as a community* out of the interconnected space that always already existed' (p. 8) and, more generally, write of 'a shared historical process that differentiates the world as it connects it' (p. 16). (Edwin Wilmsen (1989) has produced a detailed study of the places and peoples of this part of southern Africa and his argument, too, is that there is evidence of interconnectedness from more than a millennium ago (glass beads witness to contact with Asia), that received categories and 'authenticities' need to be questioned, and that the current ascriptions of remoteness and isolation have been produced, both discursively and materially, through colonialism.) All of this is now both frequently rehearsed in theory and just as frequently ignored in practice.

Gupta and Ferguson readily admit the difficulty of the project, the difficulty of wrenching ourselves out of a spatial frame to which we have so long grown accustomed. But the importance of doing it is essentially political. In a sentence which parallels, in this sphere of global cultural differentiation, Butler's arguments about personal and group identity, they write: 'The presumption that spaces are autonomous has enabled the power of topography to conceal successfully the topography of power' (p. 8).

Eric Wolf's *Europe and the people without history* (1982) has been central to all of this. Wolf's target, again, was anthropology. On the one hand, he argued, anthropology has adopted a practice of local studies and has assumed that that frame (in fact its own) relates unambiguously to the phenomena it purportedly sets out to study. And through the lens of local studies what anthropologists imagine themselves to have found are 'primitive isolates'. On the other hand, having identified these place-defined societies, argues Wolf, anthropologists have gone on to assume that they are the precapitalist 'originals'. For Wolf they are nothing of the kind. Not only are they very often precisely the product of contact through the expansion of Europe (and thus in no way 'pre' anything such as 1492), but neither is there any such thing as an 'original'. Thus: 'Everywhere in this world of 1400 [i.e. before contact with Europe], populations existed in interconnections' and 'If there were any isolated societies these were but temporary phenomena – a group pushed to the edge of a zone of interaction and left to itself for a brief moment in time. Thus, the social scientist's model of distinct and separate systems, and of a timeless "precontact" ethnographic present, does not adequately depict the situation before European expansion' (p. 71).

Both space and time are at issue here. The specificities of space are a product of interrelations – connections and disconnections – and their (combinatory) effects. Neither societies nor places are seen as having any timeless authenticity. They are, and always have been, interconnected and dynamic. As Althusser was wont to say, 'there is no point of departure'.

The modern, territorial, conceptualisation of space understands geographical difference as being constituted primarily through isolation and separation. Geographical variation is preconstituted. First the differences between places exist, and then those different places come into contact. The differences are the consequence of internal characteristics. It is an essentialist, billiard-ball view of place. It is also a tabular conceptualisation of space. It runs clearly against the injunction that space be thought of as an emergent product of relations, including those relations which establish boundaries, and where 'place' in consequence is necessarily *meeting* place, where the 'difference' of a place must be conceptualised more in the ineffable sense of the constant emergence of *uniqueness* out of (and within) the specific constellations of interrelations within which that place is set ('the impossibility of a *position* which *is not already a relation*' – Kamuf, 1991, p. xv) and of what is made of that constellation. This latter is a specificity which is elaborated by Oakes, Wolf, Wilmsen … as process, as the constant production of the new; neither an essentialised emergence from an origin nor the product of a spacing in the sense of expulsion or attempted purification; and it indicates the dubiousness of that duality – so popular and so persistent – between space and place.

Moreover, not only under modernity was space conceived as divided into bounded places but that system of differentiation was also organised in a particular way. In brief, spatial difference was convened into temporal sequence. Different 'places' were interpreted as different stages in a single temporal development. All the stories of unilinear progress, modernisation, development, the sequence of modes of production … perform this operation. Western Europe is 'advanced', other parts of the world 'some way behind', yet others are 'backward'. 'Africa' is not *different* from Western Europe, it is (just) behind. (Or maybe it is indeed only different *from*; it is not allowed its own uniqueness, its coeval existence.) That turning of the world's geography into the world's (single) history is implicit in many versions of modernist politics, from liberal progressive to some Marxist. Euphemistically to relabel 'backward' as 'developing' and so forth does nothing to alter the significance and import of the fundamental manoeuvre: that of rendering coexisting spatial heterogeneity as a single temporal series.

Now, this characteristic manoeuvre of modernity is frequently recognised, and it is a manoeuvre with clear implications. In these conceptions of singular progress (of whatever hue), temporality itself is not really open. The future is already foretold; inscribed into the story. This is therefore a temporality which anyway has none of the characteristics of event, or of novelty. Nor does it live up to the requirement that space be always and ever open, constantly in a process of being made.

The temporal convening of space thus reworks the nature of difference. Coexisting heterogeneity is rendered as (reduced to) place in the historical queue. As Sakai (1989) writes, history is 'not only temporal or chronological but also spatial and relational. The condition for the possibility of conceiving of history as a linear and evolutionary series of incidents lay in its not as yet thematized relation to other histories, other *coexisting* temporalities' (p. 106; emphasis in the original). This is an act which suppresses the full measure of the differences at issue. It is a point explored, though with a different inflection, by Johannes Fabian in relation to anthropology. For him, the crucial aspect of the manoeuvre is that anthropologists, by placing 'those who are observed' in a different time from 'the Time of the observer' (1983, p. 25) 'sanctioned an ideological process by which relations between the West and its Other, between anthropology and its object, were conceived not only as difference, but as distance in space *and* Time' (p. 147; emphasis in the original). 'Time is used to create distance in contemporary anthropology' (p. 28). Here then (i) conceptualisations of space and time (what Fabian aptly renders as 'political cosmologies') are central to the construction of a particular form of power/knowledge. Like Hall, Fabian is insisting on colonialism both as a system of rule and as a system of power/knowledge, and it is this latter aspect, of 'cognitive complicity' (p. 35), which he is mainly addressing. Moreover, (ii) the temporal convening of space is here being used to *increase* distance. Specifically, it shifts the object of study to a decent remove from the source of the scientific gaze (that this is daily contradicted by the anthropologist's practice of fieldwork, and thus of actually *talking* to this temporally distanced other is a tension (short of time travel) central to Fabian's argument). However, (iii) as in the similar strategies of modernist narratives, this greater distancing has the effect of *decreasing* the actuality (one might say the challenge) of difference. Once again what is going on here is the taming of space. The suppression of what it presents us with: actually existing multiplicity. The refusal to face up to space as quite the opposite of 'the dead, the fixed, the immobile'. The object of anthropology's gaze, as Fabian puts it, is not there and then but there and *now*, and that is a much bigger challenge.[5] Difference/heterogeneity here is not only neatly packed into its bounded spaces but also dismissed to the ('our') past. The modernist, anthropological and, as we shall see also still very much alive, temporal convening of space refuses to recognise what Fabian calls 'coevalness'. He writes 'coevalness aims at recognizing cotemporality as the condition for truly dialectical confrontation' (p. 154) and 'What are opposed … are not the same societies at different stages of development, but different societies facing each other at the same Time' (p. 155). It is important to emphasise that this radical contemporaneity does *not* imply either a romanticised/exoticised radical difference or a blandly relativistic denial that there is any such thing as 'progress', say, or 'development', at all. What may be criticised in the latter are assumptions of singularity and a lack of democracy in their determination. Coevalness concerns a stance of recognition and respect in situations of mutual implication. It is an imaginative space of engagement: it

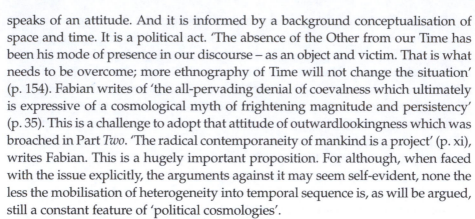
speaks of an attitude. And it is informed by a background conceptualisation of space and time. It is a political act. 'The absence of the Other from our Time has been his mode of presence in our discourse – as an object and victim. That is what needs to be overcome; more ethnography of Time will not change the situation' (p. 154). Fabian writes of 'the all-pervading denial of coevalness which ultimately is expressive of a cosmological myth of frightening magnitude and persistency' (p. 35). This is a challenge to adopt that attitude of outwardlookingness which was broached in Part *Two*. 'The radical contemporaneity of mankind is a project' (p. xi), writes Fabian. This is a hugely important proposition. For although, when faced with the issue explicitly, the arguments against it may seem self-evident, none the less the mobilisation of heterogeneity into temporal sequence is, as will be argued, still a constant feature of 'political cosmologies'.

The different aspects of this taming of the spatial are connected. The lack of openness of the future for those 'behind' in the queue is a function of the singularity of the trajectory. Ironically, not only is this temporal convening of the geography of modernity a repression of the spatial, it is also the repression of the possibility of other temporalities. The long-hegemonic temporal convening of the geography of modernity entails the repression of the possibility of other trajectories (other, that is, than the stately progress towards modernity/modernisation/development on the Euro-Western model).[6] It is a repression which can be seen as a kind of initiating counterpoint to the provocation of the *end* of modernity – if such it is – by the so-called 'arrival of the margins at the centre'. And as such it explains why this arrival, and the accompanying reassertion of the depth of the differences at issue, came as such a shock to the West. To rewrite it in Fabian's terminology, it was not merely the arrival of what have frequently been called 'the margins' (a spatial concept) but the arrival of people from *the past*. Distance was suddenly eradicated both spatially *and* temporally. Migration was thereby an assertion of coevalness. Moreover, and by the same means, the repression of the spatial was bound up with the establishment of foundational universals (and *vice versa*), the repression of the possibility of multiple trajectories, and the denial of the real difference of others. In a whole variety of ways, what was at issue was the establishment of a geography of power/knowledge. Yet it was also a deeply ironic one, for what it entailed was the suppression of the real challenges of space.

There is one further twist. In the previous chapter the rather odd notion was explored that space conquers time. It is assumed to do this, I suggested, through the equally assumed equation of space with representation. Spatialisation, in the guise of the writing down of the temporal, conquers time. It takes the life out of the essentially temporal world. (My argument in reply was that the mistaken move here was the equation of representation with space. That while representing time might take the life out of time, equating representation with space takes the life out of space. We have a whole cemetery of dimensions on our hands.) Moreover, and precisely as a result of this formulation, it is frequently averred that the opposite cannot happen: space may conquer time but time cannot conquer space: 'the opposite is not possible: time cannot hegemonize anything' (Laclau, 1990, p. 42).

Yet the opposite *has* happened, and continues to happen, and with significant effects. In many of these discourses of modernity contemporaneous differences have been conceptualised as temporal sequence.[7] The multiplicities of the spatial have been rendered as merely stages in the temporal queue. It is a discursive victory of time over space. (Of course, it would still be possible for the intransigent to maintain that there was no contradiction here, that representation *as such* is still spatialisation – it just happens that this particular representation mobilised time to represent space – Kern (1983) effectively resorts to this. The tortured complexity of this argument indicates the difficulty with the initial equation of representation with the spatial.) This, then, is quite the opposite of the usual view. Here the representation of *space* takes place through its convening into a temporal sequence. The challenge of space is addressed by an imagination of time. In these discourses of modernity there was *one* story, which the 'advanced' countries/peoples/cultures were leading. There was only one history. The real import of spatiality, the possibility of multiple narratives, was lost. The regulation of the world into a single trajectory, *via* the temporal convening of space, was, and still often is, a way of refusing to address the essential multiplicity of the spatial. It is the imposition of a single universal.

This kind of space of modernity, in other words, doesn't see space as emerging from interaction, nor as the sphere of multiplicity, nor as essentially open and ongoing. It is a taming of the challenge of the spatial. This is a far deeper victory of time over space than the oft-referred-to deprioritisation. 'Recognising spatiality' involves (could involve) recognising coevalness, the existence of trajectories which have at least some degree of autonomy from each other (which are not simply alignable into one linear story). It is this that I shall take it to mean in what follows. On this reading, the spatial, crucially, is the realm of the configuration of potentially dissonant (or concordant) narratives. Places, rather than being locations of coherence, become the foci of the meeting and the nonmeeting of the previously unrelated and thus integral to the generation of novelty. The spatial in its role of bringing distinct temporalities into new configurations sets off new social processes. And in turn, this emphasises the nature of narratives, of time itself, as being not about the unfolding of some internalised story (some already-established identities) – the self-producing story of Europe – but about interaction and *the process of the constitution of* identities – the reformulated notion of (the multiplicities of) colonisation.

(A reliance on science? 2)

The modernist conception of nation-states or cultural isolates resonates with the billiard-ball view of the world proposed by physical mechanics. First the entities exist, in their full identities, and then they come into interaction. There is a distinct inside and outside. It is a useful analogy. The move towards relational identities, openended futures and such-like, can similarly be read as analogous to subsequent developments in natural sciences.

There are many who have made this move, and my doubts arise only where the parallels seem to be assumed to be far more than provocative analogies. The dubiousness of attempts to have recourse to the natural sciences as any form of ultimate legitimation has already been discussed in Part Two. (The reverential reference: 'It must be right because physics says so' etc.) It is unstable ground on which to rest one's case. It is rare that one can unequivocally appeal to, say, 'recent developments in physics' in proof or demonstration of an argument in another field, for such developments are often themselves the subject of fierce debate. Think for instance of the wrangles over quantum theory, or evolution. Given the kind of imagination of space that I am proposing I could easily appeal to witnesses in some branch of the natural sciences in corroboration of my argument. But I could also – being honest – find a bunch of natural scientists who propose quite a contrary point of view. And, within the natural sciences, I am not competent to judge. Perhaps, therefore, we ought not to resort to tactics that in reality amount to picking out for quotation one's favourite, or most compatible, 'harder' scientist.

It is, moreover, somewhat sobering to consider previous attempts to adopt this strategy. Presumably those enthusiastically following earlier scientists were as confident and excited as are the exponents and adopters of the likes of complexity theory today. Yet consider what Fabian has to say about the roots of modernist political cosmology (he is mainly considering time) in a combination of the then new evolutionary sciences and 'Newtonian physicalism':

> The use of Time in evolutionary anthropology, modeled on that of natural history, undoubtedly was a step beyond premodern conceptions. But it can now be argued that wholesale adoption of models (and of their rhetorical expressions in anthropological discourse) from physics and geology was, for a science of man, sadly regressive intellectually, and quite reactionary politically. (1983, p. 16)

Having spelled out what he sees as some of the regressive implications, he observes:

> This was politically all the more reactionary because it pretended to rest on strictly scientific hence universally valid principles. (p. 17)

Perhaps, too, in the case of space, the scientific legitimacy of an atomistic imagination has been of critical importance in providing a background to a cosmology of an essentially regionalised space, to claims for the essential belongingness of a people with its place, for the necessity of boundaries against incursions from an essentially foreign outside, for the innumerable telluric myths of origins, … and so on and so forth.

Fabian suggests a further possible political repercussion of this reliance on science, one which takes us back both to the temporal convening of spatial difference and, yet again, to the meeting of Moctezuma and Cortés. He has been, at this point, considering the idea of 'Physical Time':

> In the hands of ideologues such a time concept is easily transformed into a kind of political physics. After all, it is not difficult to transpose from physics to politics one of the most ancient rules which states that it is impossible for two bodies to occupy the same space at the same time. When in the course of colonial expansion a Western body politic came to occupy, literally, the space of an autochthonous body, several alternatives were conceived to deal with that violation of the rule. The simplest one, if we think of North America and Australia, was of course to move or remove the other body. Another one is to pretend that space is being divided and allocated to separate bodies. South Africa's rulers cling to that solution [this was published in 1983]. Most often the preferred strategy has been simply to manipulate the other variable – Time. With the help of various devices of sequencing and distancing one assigns to the conquered populations a different Time. (1983, pp. 29–30; emphasis in the original)

This is in no way to argue against talking between fields (Massey, 1996b). But it is to urge caution and, most importantly, an explicit awareness of the terms of the conversation. In the light of this history there is a need to be wary about the current fascination with complexity theory, fractals, quantum mechanics and the rest. Not only might this version of things, like previous ones, fade or become just a part of the story, but also we need to be radically aware of its potential political implications. There are many who are now haughtily critical of many previous readings. Those who adopt what Robbins sees as 'the unreflective scorn for modernity among Western intellectuals' (1999, p. 112) should be aware that the same dismissal may await their own position a generation or so down the line. One of Fabian's criticisms of anthropology's strategy (the way in which it was 'regressive intellectually') is that it was, in its reliance on science, simply out of date: 'anthropology achieved its scientific respectability by adopting an essentially Newtonian physicalism (…) at a moment near the end of the nineteenth century when the outlines of post-Newtonian physics (…) were clearly visible' (p. 16). Those postmodern writers in social science and the humanities who today rest their case, with the same degree of enthusiasm, on 'the new sciences' should both beware of this history and also remember that unreflective acceptance, as opposed to active engagement, was precisely the kind of strategy which that wonderfully nomad philosopher Henri Bergson did not adopt.

(Representation again, and geographies of knowledge production 1)

The era of classical science was also associated with a dominant conception of certain aspects of what might be called the geography of knowledge production. And, again, these characteristics were mimicked by a social science in awe of its neighbour across the campus. Isabelle Stengers (1997) recounts in detail the choice that physicists made, as she puts it, between Einstein and Kepler. They chose Einstein, and with him an understanding of physics as being concerned with 'fundamental laws'. Fundamental laws as opposed to the 'merely phenomenological', the messiness of 'the real world'. They also decided, moreover, that all things – including those messy phenomenological things – were in the end accountable for by the fundamental laws (any current inability actually to carry this off was ascribed to the fact that science hadn't got there 'yet'). By the end of the nineteenth century, however (and the work of Ludwig Boltzmann is classically cited here as of particular significance), this formulation was already coming up against the problem of time … 'physicists realised that the laws they had taken for granted for about two centuries and accepted as fundamental did not allow them to distinguish between before and after!' (Stengers, p. 23). And so began the fierce controversies referred to in Part Two. But what is relevant here is that this option for fundamental laws represented an understanding of science as a particular form of abstraction from the merely phenomenal 'real world'. The form of the gap is what is significant: those laws were removed from embodiment and encapsulated into language, code, equations, representations, which were then taken to be the source. N. Katherine Hayles calls it the Platonic backhand: 'The Platonic backhand works by inferring from the world's noisy multiplicity a simplified abstraction. So far so good: this is what theorizing should do. The problem comes when the move circles around to constitute the abstraction as the originary form from which the world's multiplicity derives' (1999, p. 12).

There are other kinds of gaps as well. When we convene spatial differences into temporal sequence, as did/do so many modernist narratives, we are repressing the actuality of those differences. But there is another process also going on. For Fabian, and for many others, the crucial point is that that manoeuvre articulates the knowledge relation. It instates a geography (as well as Fabian's temporality) of the production of knowledge. It is an act of distancing; the creation of a particular kind of gap. The primary aspect of this is that the process of becoming a producer of knowledge (and a definer and guardian of the kinds of things said to be knowledge) involves setting oneself apart from the things one is studying. As Fabian points out, anthropology's manoeuvrings to distance itself further from its object of study were/are not peculiar to that discipline: 'After all, we only seem to be doing what other sciences exercise: keeping subject and object apart' (1983, p. xii) – maintaining a distance between 'knower' (so-called) and 'known' (ditto). It is a separation which may – as in the case here – be produced conceptually (here by removing the known to another time). But it can also be produced materially. From the desert fathers (Waddell, 1987) through the various specialised (read: exclusive and excluding) places of Western knowledge production – the monasteries, the early universities (and some would argue many of today's univer- sities) – to the new elite locations such as science parks and Silicon Valley – there has been a social geography of knowledge production (elite; historically largely male) which gained (and continues to gain) at least a part of its prestige from the cachet and exclu- sivity of its spatiality (Massey, 1997b; Massey et al., 1992). Physical location itself has mirrored and reinforced the structure of knowledge production being carried on within it (Massey, 1995b). Moreover, the spatial structures of knowledge production which assume a radical gap between knower and known are precisely ones through which the equation between representation and spatialisation can be confirmed.

The particular way in which Fabian interprets this as happening within anthropology is through the construction of knowledge through taxonomies. Others have made a similar point in a more general context. It is through the construction of taxonomies (via distancing and visualisation) that representation through mapping, ordering, writing is made possible. Fabian writes frequently of taxonomic space (or, in relation to structuralism, 'tabular' space, after Foucault) and he distinguishes it from eco- logical space, or 'real space, perhaps the space of the human geographers' (p. 54). The pity of it is that the reputation of the former has rubbed off on the latter.

The linking of all these distinct manipulations of the term space leads to some suggestive possibilities, hinted at in Part Two. The geography of knowledge production is intimately related to the question of what is understood by representation (Latour, 1999b). Thus Fabian, among many others, urges: 'What must be developed are the elements of a processual and materialist theory apt to counteract the hegemony of taxonomic and representational approaches which are identified as the principal sources of anthropology's allochronic orientation' (1983, p. 156; emphasis in the original).[8] And what Stengers is searching for is a science which rejects the binary of fundamental- phenomenal, one which takes seriously temporal irreversibility (and indeterminacy) – 'the physics of processes cannot be reduced to a physics of states' (1997, p. 65) – and one which, although very definitely a specific form of practice, is explicitly socially embed- ded. Thrift (1996), among others, has been trying to work towards non-representational theory in geography. Perhaps these moves in the implicit spatialities of the knowledge relation might further aid in the liberation of 'space' from its old associations. And then maybe we could turn, instead, towards that much more awkward, intractable and chal- lenging thing 'real space, the space of the human geographers'. And one thing which might immediately occur to us there is the need to ponder the elitist, exclusivist, enclo- sures within which so much of the production of what is defined as legitimate knowledge still goes on.

7
instantaneity/depthlessness

We live, some say, in spatial times. There is an imagination of globalisation which pictures it as a totally integrated world. From a world structured and preoccupied by history we have landed ourselves in a depthless horizontality of immediate connections. A world, it is said, which is purely spatial. (With a delicious irony, Grossberg argues that even this assertion of the reprioritisation of space is still in thrall to temporality. This 'strategy chronologises space: for example, reprivileging history as the agency which has replaced history with geography. This is the strategy of most so-called "post-modernisms"' (1996, p. 177). Even more ironically, one might add that this is a formulation which deals in a singular history.)

In its most extreme form this view of the current state of things is an imagination of instantaneity – of a single global present. It figures in a multitude of ways: in global media events – the death of Princess Diana, the Olympic Games or the event of Tien-an-Men Square; it figures in talk of the global village, and perhaps in the propositions of an easy multiculturalism-across-the-continents in a host of advertising strategies. The extreme of instantaneity recalls, once again and in new guise, space as the seamless coherence of a structuralist structure, the essential section of a slice through time. In this formulation temporality becomes impossible – how to pass between a series of self-contained presents? History becomes unthinkable. Hence the apprehension of depthlessness. This, however, is to posit two mutually exclusive alternatives – an appreciation of the temporal and a consciousness of the instantaneous connectivity of space. They are taken, not simply as empirically mutually exclusive, but as definitionally counterposed. Instantaneity is spatial, and therefore cannot be temporal (we have come across this leap before). Once again, this is to fail to imagine the interconnectivity of the spatial as not between static things but between movements, between a plurality of trajectories. That 'the new depthlessness' poses problems for thinking historically is without doubt. But it also poses problems for thinking spatially. Just as time cannot adequately be conceptualised without a recognition of the (spatial) multiplicities through which it is generated so space cannot adequately be imagined as the stasis of a depthless, totally interconnected, instantaneity. Any assumption of a closed instantaneity not only denies space this essential character of itself constantly becoming, it also denies time its own possibility of complexity/multiplicity. To read interconnectivity as the instantaneity of a closed surface (the prison house of synchrony) is precisely to ignore the possibility of a multiplicity of trajectories/ temporalities. If this is the imagination which is to replace modernism's temporal alignment of regions then it is a move straight through from a billiard-ball world of essentialised places to a claustrophobic holism in which everything everywhere is already connected to everywhere else. And once again it leaves no opening for an active politics.

There is, of course, no single integrated global moment. McKenzie Wark's (1994) analysis of global media events demonstrates the complex, uneven and spatially differentiated nature of their construction (and the emphasis on construction is important). The heterogeneous nature of the world's articulation into these temporary time-space constellations serves to highlight, rather than to indicate the elimination of, the significance of multiplicity. Indeed, the construction of these media events *as* global is precisely an outcome of the intersections within such a multiplicity. They are constructed 'places' of virtual geographies:

> An urban site redolent with symbolic meaning; a panoptic political regime struggling to contain its own power in the face of a modernity it both ardently desires and resolutely opposes; the presence of the Western media with their global information vectors: Tienanmen Square in April, May, and June of 1989 was a metaphorical crossroads for the intersection of diverse forces, following different trajectories at different speeds. In Lenin's terms it formed a conjuncture; in Althusser's, a point of overdetermination. (p. 127)

And anyway, the understanding of globalisation as an achieved instantaneity is ambiguous from the off. On the one hand it is often, at least implicitly, claimed to be already with us. On the other hand it is the very promise of a future-to-come which globalisation is said to hold out. It is this latter proposition which allows those who are not 'yet' integrated into this single globality to be figured as backward, as still temporally 'behind'. In this double formulation the singular temporality which is the assumption of the convening of spatial difference into temporal sequence will find its consummation in the single temporality of a unified global present.

It is precisely this shift, from vertical to horizontal if you like, which is argued by Fredric Jameson (1991) to characterise the movement from the modern to the postmodern. While during the modern period the very survival of 'nature', of 'the traditional countryside and of traditional agriculture' (p. 311), that is, of 'uneven development' itself (p. 366), provided the conditions for an idea of historicity, of the new and indeed of the notion of 'eras' at all, with the advent of the 'late capitalism' which Jameson sees as the economic foundation of the postmodern:

modernization triumphs and wipes the old completely out: nature is abolished along with the traditional countryside and traditional agriculture; even the surviving historical monuments, now all cleaned up, become glittering simulacra of the past, and not its survival. Now everything is new; but by the same token, the very category of the new then loses its meaning ... (p. 311)

Regardless of the empirical basis of this claim it is important to note its conceptual foundation. Under Jameson's reading of the modern, actually existing differences, such as uneven development, are characterised temporally: they are residues, they lend 'us' a notion of history (of where we are coming from) and, correlatively, of the new and of the future. There is only one trajectory here. Under his reading of *post*modernity, because the laggards have now caught up or been obliterated or simulacralised we are all in a single time, which is the present, a condition which in turn makes it impossible for us to have a sense of temporality, of history, at all:

the postmodern must be characterized as a situation in which the survival, the residue, the holdover, the archaic, has finally been swept away without a trace. In the postmodern, then, the past itself has disappeared (along with the well-known 'sense of the past' or historicity and collective memory). ... Ours is a more homogeneously modernized condition; we no longer are encumbered with the embarrassment of non-simultaneities and non-synchronicities. Everything has reached the same hour on the great clock of development or rationalization (at least from the perspective of the 'West'). (pp. 309–10)

While I would not want to quarrel with Jameson's diagnosis of postmodern (or modern) political cosmologies, it is important to pull out what is going here. This a-temporal single time is called by Jameson 'space': 'So, even if everything is spatial, this postmodern reality here is somehow more spatial than everything else' (p. 365). This is space as stasis, as equated with depthlessness.

Jameson also counterposes space as a closed synchrony (the postmodern) to space as convened into a single temporal linearity (the modern). In my view neither of them is an adequate formulation of space or of time. Jameson's response to a depthless world, as he sees it, is to replace it with one where depth takes the form of a single history, which organises spatial difference. We do, certainly, need a new imagination but a return to that regionalising, temporally convening, one of modernity does not provide a politically adequate alternative. The shift in viewpoint, so common in comparisons of modernity and postmodernity, from *one* history to *no* histories, from a single (progressive) story to a synchronic depthlessness, in both eras though in radically different ways, denies the real challenge of the spatial.

But Jameson's reasons for this manoeuvre, his desire to return to a single ordering history, are also important to note. For him, multiplicity can provoke terror. For Jameson, if we do not understand the world in terms of some cultural dominant 'then we fall back into a view of present history as sheer heterogeneity, random difference, a coexistence of a host of distinct forces whose effectivity is undecidable' (p. 6) (hang on: why does heterogeneity have to be sheer, or difference random, or the lack of a single dominating force render everything undecidable?); it leaves us with 'the messiness of a dispersed existence' (p. 117) and – that other aspect of a shift away from modern spatiality – 'the strange new feeling of an absence of inside and outside' (p. 117) '… the security of the Newtonian earth withdrawn' (p. 116).

However, while the terms of his response may be disputed, what Jameson is here certainly alive to is aspects of the challenge of a full recognition of the spatial. And indeed, one especially fascinating element of his analysis is the link he makes between the new consciousness of this massive heterogeneity and what he calls 'the demographies of the postmodern' (p. 356). In some wonderful passages he writes that 'The West … has the impression that without much warning and unexpectedly it now confronts a range of genuine individual and collective subjects who were not there before' (p. 356) and of 'some new visibility of the "others" themselves, who occupy their own stage – a kind of centre in its own right – and compel attention by virtue of their voice and of the act of speaking itself' (p. 357). Here are brought together: international migration (from a specifically Western point of view), the end of modernity, and the assertion of coevalness.[9] For Jameson, who recognises the ethnocentricity and racism within all this, it is these huge movements which ground the shift in perspective on the part of those who get to tell the stories of 'our times'.

He cites Sartre trying to come to grips, in the very moment of his own thinking, with the fact of Communists and Nazis fighting in Berlin, unemployed workers marching in New York, 'boats on the open sea that are echoing with music', and lights 'going on in all the cities of Europe' (Sartre, 1981, p. 67, cited in Jameson, 1991, pp. 361–2). Jameson rates this passage of Sartre as 'pseudoexperience', 'as a failure to achieve representation', as 'voluntaristic, an assault of the will on what is "by definition" structurally impossible of achievement rather than something pragmatic and practical that seeks to augment my information about the here and now' (all p. 362). 'It seems at the same time to be a relatively aimless and exploratory fantasy as well, as though the subject were afraid of forgetting something but could not quite imagine the consequences: Will I be punished if I forget all the others busy living simultaneously with me?' (p. 362). Now, at one level it is clear what Jameson means: the passage from Sartre is evocative (though for me productively evocative) and not analytical. But it is meant to be. Jameson's complaint at the 'failure to achieve representation' seems to refer to the inevitable incompleteness of content (what has been left out?). Is this an implicit claim by Jameson that (complete) representation *was* possible when we didn't have to deal with all this confusing coevalness? (When we could pull everything into shape under the tutelage of the one narrative of the period in dominance? When convening space into temporal sequence enabled its representation?) It is this kind of 'representation' which denies the multiplicity of the spatial.

Jameson, though, does have a real point. The difficulty of representing the spatial ('a simultaneity of distinct streams of elements which the senses grasp altogether', p. 86) is something he returns to again and again. It is a reading opposite to that of Laclau. For Laclau space was, precisely, the closure *of* representation. For Jameson the reality of the spatial is its very unrepresentability.[10] To associate this only with postmodernism, however, would be to acquiesce in that reading of modernity in which contemporaneous heterogeneity is representable (and thereby its challenge, both to representation and politically, obliterated) through its reduction to temporal sequence: as we have seen, to recognise the spatiality of modernity would make that 'era' a challenge to representation in that sense too. But the underlying point catches at something significant: that far from standing for the stability *of* representation, real space (space-time) is indeed impossible to pin down.

But anyway, the argument should not really be about content (some patently vain attempt, in an evocation of a simultaneity of stories-so-far, to enumerate each and every one of those trajectories). Rather, it is a question of the angle of vision, a recognition of the *fact* (not all of the content) of other realities, equally 'present' though with their own histories. Of course we cannot recount them all, or be constantly aware of each and every one of 'the others busy living simultaneously with me'. Perhaps what is needed first is a leap into space. Then there will be a prioritisation, a selection, perhaps reflecting actual practices of relationality. Perhaps it is apposite here to recall Grosz's arguments about subjectivity. Perhaps what is required is the inculcation of a (notion of) subjectivity which is not exclusively temporal; not the projection of an interior – conceptual, introspective (see Part *Two*), but rather a subjectivity which is spatial too, outwardlooking in its perspectives and in the awareness of its own relational constitution.

8
aspatial globalisation

'Globalisation' is currently one of the most frequently used and most powerful terms in our geographical and social imaginations. At its extreme (and though extreme this version is none the less highly popular) what it calls up is a vision of total unfettered mobility; of free unbounded space. In spite of searching and provocative interventions from the likes of Anthony King, Jan Niedeven Pieterse, Michael Peter Smith, Arjun Appadurai and many others, this vision persists. In academic work, it perhaps finds its most characteristic presence as a summary of economic globalisation in the opening paragraphs to a treatise on something 'more cultural'. But it is an understanding which also thoroughly permeates popular, political and journalistic discourse. At its worst, it has become a mantra. Characteristic words and phrases make an obligatory appearance: instantaneous; Internet; 24-hour financial trading; the margins invading the centre; the collapse of spatial barriers; the annihilation of space by time. In these texts, the emerging world economy will be captured by an iconic economics: reference to CNN, McDonald's, Sony is frequently considered enough to convey it. And judicious alliterations will strive to convey the maziness of it all: Beijing – Bombay – Bamako – Burnley. What are at issue in all of this are our geographical imaginations. (And in this regard the alliterations are of particular interest: how often they reveal, in their expectations of the effects they will produce, an imaginative geography which still knows which is 'the exotic' and which 'the banal' and when it is bringing them into unexpected (though in fact now so common a trope) juxtaposition.) It is a mantra which evokes a powerful vision of an immense, unstructured, free unbounded space and of a glorious, complex mixity.[11]

It is also, undoubtedly, an imagination of the world's geography (a political cosmology in Fabian's terms) which contrasts radically with the modernist one. In place of an imagination of a world of bounded places we are now presented with a world of flows. Instead of isolated identities, an understanding of the spatial as relational through connections. The very word 'globalisation' implies a recognition of spatiality. It is a vision which in some sense glorifies (as so much current writing does) in the triumph of the spatial (while at the same time speaking of its annihilation). Yet if the picture of global space which 'globalisation' evokes is in contrast to the dominant imaginary under modernity, the structuring characteristics of the conceptualisation of space are disarmingly similar.

Most obviously, just as in the old story of modernity, this is a tale of inevitability; and this in turn is enabled by an unspoken concept of space. Clinton's analogy with the force of gravity only highlights in a particularly striking way what is routinely taken for granted. Whether through an unthinking technological determinism or through a submission to the inevitability of market expansion, this version of globalisation comes to have almost the ineluctability of a grand narrative. Globalisation, here, is as inevitable as modernity's story of progress, and the implications, again, are enormous. Yet again, and just as in modernity's discourse, spatial differences are convened under the sign of temporal sequence. Mali and Chad are not 'yet' drawn into the global community of instantaneous communication? Don't worry; they soon will be. Soon they will, in this regard, be like 'us'.

This is an *aspatial* view of globalisation. The potential differences of Mali's and Chad's trajectories are occluded. (The essential multiplicities of the spatial are denied.) Such countries are assumed to be following the same ('our') path of development. (The openness of the future which is in part a consequence of the multiplicities of the spatial is reined in. This is a tale with a single trajectory.) The effects are political. Because space has been marshalled under the sign of time, these countries have no space – precisely – to tell different stories, to follow another path. They are dragooned into line behind those who designed the queue. Moreover, not only is their future thus supposedly foretold but even this is not true, for precisely their entanglement within the unequal relations of capitalist globalisation ensures that they do not 'follow'. The future which is held out as inevitable is unlikely to be reached. This convening of contemporaneous geographical differences into temporal sequence, this turning it into a story of 'catching up', occludes present-day relations and practices and their relentless *production*, within *current* rounds of capitalist globalisation, of increasing inequality. It occludes the power-geometries within the contemporaneity of *today's* form of globalisation. Even within the West, European governments following the US model appeal to the 'future' in justification, thereby closing down a politics in which a European approach might challenge that of the USA. As Bruno Latour has written, 'Just at the moment when there is much talk on the topic of globalisation, it is just the time *not* to believe that the future and the past of the United States are the future and the past of Europe. A left party should produce a new difference' (1999a, p. 14).

It is, further, significant that such tales of inevitability require dynamics which are beyond intervention. They need an external agent, a *deus ex machina*. The unquestioned motors of 'globalisation's' historicising of the world's geographical inequalities are, in various mixtures, the economy and technology. By this means, a further political result is achieved: the removal of the economic and the technological from political consideration. The only political questions become ones concerning our subsequent adaptation to their inevitability. Latour (1999a) has written powerfully of this widespread move to protect 'the economic' – that is, the capitalist market – from political questioning (he writes also of an equivalent move in relation to Science). All this has as a necessary grounding the conversion of space into time: the consequent occlusion of the contemporaneous multiplicity of the spatial occludes also the nature of the relations in play.

Further, the particular form of globalisation which we are experiencing at the moment (neoliberal capitalist, led by multinationals, etc. etc.) is taken to be the one and only form. Objections to this particular globalisation are persistently met with the derisive riposte that 'the world will inevitably become more interconnected'. Capitalist globalisation is equated with globalisation *tout court*, a discursive manoeuvre which at a stroke obscures the possibility of seeing alternative forms. It is globalisation *in this particular form* which is thereby taken as being inevitable. The 'achievement' here is to make into the political stake an abstract spatial scale ('the global'), and incidentally to stimulate a response which defends 'the local'. It is, rather, the relations which mutually construct them both which need to be the object of dispute.

Finally, that way of seeing globalisation as inevitable, of placing economics/technology beyond the reach of political debate, also renders globalisation as the One story. 'Globalisation', just as the term 'Capitalism' was before it (and for which, as did modernity in its own day, it frequently stands in as an obfuscating euphemism), is the one (self-referential) Identity in relation to which all else is defined (see Gibson-Graham, 1996). That, again, is to fail to recognise the multiplicities of the spatial. Globalisation is not a single all-embracing movement (nor should it be imagined as some outward spread from the West and other centres of economic power across a passive surface of 'space'). It is a making of space(s), an active reconfiguration and meeting-up through practices and relations of a multitude of trajectories, and it is there that lies the politics.

The imagination of globalisation in terms of unbounded free space, that powerful rhetoric of neoliberalism around 'free trade', just as was modernity's view of space, is a pivotal element in an overweaning political discourse. It is a discourse which is dominantly produced in the countries of the world's North (though acquiesced in by many a government in the South). It has its institutions and its professionals. It is normative; and it has effects.

In the South it is this understanding of the space of the future (as unbounded global trading space) which enables the imposition of programmes of structural adjustment, and their successors. It is this understanding of the unavoidability of this form of globalisation which legitimises the enforcement of export orientation on the economy of country after country; the prioritisation of exports over production for local consumption. It is this discourse of, this particular form of, globalisation in other words which is an important component in the continuing legitimisation of the view that there is one particular model of 'development', one path to one form of 'modernisation'.

In the North, too, this geographical imagination has effects: the constant talking about it, the endless describing it in a particular form, is part of the active project of its production. It becomes the basis for decisions precisely to implement it. On the one hand globalisation is represented as ineluctable – a force in the face of which we must adapt or be cast into oblivion. On the other hand some of the most powerful agencies in the world are utterly intent on its production. The duplicity of the powerful in this is deep, and has been characterised by Morris (1992b) in terms of eroticism (see also, for an alternatively ribald account, Lapham, 1998). World economic leaders gather (in Washington, Paris or Davos) to congratulate themselves upon, and to flaunt and reinforce, their powerfulness, a powerfulness which consists in insisting on power*lessness* – in the face of globalising market forces there is absolutely nothing that can be done. Except, of course, to push the process further. It is a heroic impotence, which serves to disguise the fact that this is really a *project*.

This vision of global space, then, is not so much a description of how the world is, as an image in which the world is being made. Just as in the case of modernity, here we have a powerful imaginative geography. It is a very different imagination: instead of space divided-up and bounded here is a vision of space as barrier-less and open. But both of them function as images in which the world is made. Both of them are imaginative geographies which *legitimise* their own production.

Clearly, the world is not totally globalised (whatever that might mean); the very fact that some are striving so hard to make it so is evidence of the project's incompletion. But this is more than a question of incompletion – more than a question of waiting for the laggards to catch up. There are multiple trajectories/ temporalities here. Once again, as in the case of modernity, this is a geographical imagination which ignores the structured divides, the necessary ruptures and inequalities, the exclusions, on which the successful prosecution of the project itself depends. A further effect of the temporal convening of spatial difference here again becomes evident. So long as inequality is read in terms of stages of advance and backwardness not only are alternative stories disallowed but also the fact of the *production* of poverty and polarisation within and through 'globalisation' itself can be erased from view. This is – again – a geographical imagination which ignores its own real spatiality.

Forget, for a moment, Sony and CNN. An alternative iconic economics will tell a tale of the *production* of inequality, division and exclusion. Like the old story of modernity, the new hegemonic tale of globalisation is told as a universal story, but the process is one which is not (and on current terms cannot be) universalised.

The debate about globalisation is often asserted to be about how new it is and how far it has progressed, and there clearly *is* argument about this. There are 'hyperglobalisers' such as Ohmae (1994). And there are sceptics. Hirst and Thompson (1996a, 1996b), for instance, argue that the major world national economies are no more open in terms of trade or capital flows than they were in the period of the Gold Standard. They point out that over the medium term (say the last century), there has been no monotonic linear direction of change. Instead, the degrees of openness have fluctuated over time with the nature of economic development. Their argument is well taken. However, to restrict the argument to this matter of the *degree* of globalisation is gravely to impoverish it. What should be at issue is also the *form* of globalisation: the social form of the relationality which structures it. There may be disagreements over the changes in the degree of openness of national economies over the period studied by Hirst and Thompson (and much squabbling over the details of which measures are the most appropriate), but what surely cannot be in doubt is that the world geography of those relations has been transformed. Global space, as space more generally, is a product of material practices of power. What is at issue is not just openness and closure or the 'length' of the connections through which we, or finance capital, or whatever … go about our business. What are at issue are the constantly-being-produced new geometries of power, the shifting geographies of power-relations. The *meaning* of economic openness to, say, the UK at the start of the twentieth century, with the country still clinging on to its imperial pomp and this the high point of the Gold Standard, is quite different from its meaning now, with the country's dependence on foreign inward investment and, after the ravages of the 1980s on its production of the means of production, its need to bring in from elsewhere so many of the tools of its trade. In the earlier period 'openness' spoke of dominance; the openness of today is far more ambiguous. The reluctance to address the changing form of globalisation over time is on a par with, and reinforces, the blindness to the possibility that it could take different forms *now*. Space – here global space – is about contemporaneity (rather than temporal convening), it is about openness (rather than inevitability) and it is also about relations, fractures, discontinuities, practices of engagement. And this instrinsic relationality of the spatial is not just a matter of lines on a map; it is a cartography of power.

All of which raises a final source of concern about this formulation of globalisation. It returns us again to the discursive strategies of free market (so-called) globalisation. The dominant institutions and governments which clamour most strongly in favour of globalisation argue for it in terms of free trade. And they argue for 'free trade' in terms which in turn suggest that there is some self-evident right

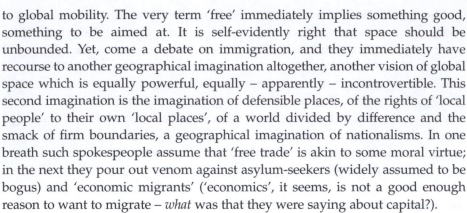

to global mobility. The very term 'free' immediately implies something good, something to be aimed at. It is self-evidently right that space should be unbounded. Yet, come a debate on immigration, and they immediately have recourse to another geographical imagination altogether, another vision of global space which is equally powerful, equally – apparently – incontrovertible. This second imagination is the imagination of defensible places, of the rights of 'local people' to their own 'local places', of a world divided by difference and the smack of firm boundaries, a geographical imagination of nationalisms. In one breath such spokespeople assume that 'free trade' is akin to some moral virtue; in the next they pour out venom against asylum-seekers (widely assumed to be bogus) and 'economic migrants' ('economics', it seems, is not a good enough reason to want to migrate – *what* was that they were saying about capital?).

Hélène Pellerin (1999) has analysed the shift from embedded liberalism to neoliberalism, and the different spatial settlements involved in each. As she points out, neoliberalism in practice is *not* simply about mobility: it too requires some spatial fixes. And of singular importance among them is the spatial organisation of labour. (And just as the imposition of free trade is contested so too is the attempt to engineer a new geography of labour – in particular she points to illegal migration flows and to aboriginal alliances.)

So here we have two apparently self-evident truths, a geography of borderlessness and mobility, and a geography of border discipline; two completely antinomic geographical imaginations of global space, which are called upon in turn. No matter that they contradict each other; because it works. And it 'works' for a whole set of reasons. First, because each self-evident truth is presented separately. But second, because while neither imagination in its pure form is possible (neither a space hermetically closed into territories nor a space composed solely of flows) what is really needed politically is for this tension to be negotiated explicitly and in each specific situation. This parallels the structure of Derrida's (2001) argument on hospitality. Each 'pure' imagination on its own tames the spatial. It is their *negotiation* which brings the question (rights of movement/rights of containment) into politics. The appeal to an imagination of pure boundedness or pure flow as self-evident foundation is neither possible in principle nor open to political debate.

And so in this era of 'globalisation' we have sniffer dogs to detect people hiding in the holds of boats, people dying in the attempt to cross frontiers, people precisely trying to 'seek out the best opportunities'. That double imaginary, *in the very fact of its doubleness,* of the freedom of space on the one hand and the 'right to one's own place' on the other, works in favour of the already-powerful. Capital, the rich, the skilled … can move easily about the world, as investment, or trade, as sought-after labour or as tourists; and at the same time, whether it be in the immigration-controlled countries of the West, or the gated communities of the rich in any major metropolis anywhere, or in the elite enclosures of knowledge production and high technology, they can protect their fortress homes. Meanwhile the poor and the unskilled from the so-called margins of this world are both instructed to open up their borders and welcome the West's invasion in whatever form it comes, and told to stay where they are.

Once again there are echoes here of how the story of modernity was told. Just as was Toussaint l'Ouverture's claim to participate in the principles of modernity's legitimating discourse, so too today the claim to free mobility (the discourse of globalisation) by the world's poor is rejected out of hand. (Though – as with the Haitian slaves – the proclamation of 'free trade' has made the challenge possible.) The current world order of capital's (anyway highly unequal) globalisation is as predicated upon holding (some kinds of) labour in place as was early modernity upon slavery. Pellerin's account of the bullying disdain with which the US government treated the issue of Mexican migration during the negotiation of NAFTA reminds one of nothing so much as C.L.R. James' account of the Parisian reply to the claims of Toussaint l'Ouverture. If, in Bhabha's words, the discourse of modernity fuelled 'the archaic racial factor in the society of slavery' (1994, p. 244) (although of course it was anything but archaic), then, too, the discourse of globalisation as free movement about the world is fuelling the 'archaic' (but not) sentiments of parochialism, nationalism and the exclusion of those who are different.

Today's hegemonic story of globalisation, then, relates a globalisation of a very particular form. And integral to its achievement is the mobilisation of powerful (inconsistent, falsely self-evident, never universalisable – but powerful) imaginations of space.

How easy it is to slip into ways of thinking that repress the challenge of space; and how politically significant spatial imaginaries can be. 'Globalisation', told in this way, is like the old story of modernity. Once again it convenes spatial difference into temporal sequence, and thereby denies the possibility of multiple trajectories; the future is not held open. This rendering of globalisation provides the framing inevitability for the construction of politics such as the 'Third Way' with its abolition of Left and Right and its political closure around a discourse which doesn't allow for dislocation – what Chantal Mouffe has called 'a politics without adversaries' (1998). It installs an understanding of space, the 'space of flows', which, just like the space of places of modernity, is deployed (when needed) as a legitimation for its own production and which pretends to a universality which anyway in practice it systematically denies. For, in fact, in the context of and as part of this 'globalisation' new enclosures are right now being erected.

And, just like the old story of modernity too, this imagination of globalisation is resolutely unaware of its own speaking position: neoliberal to be sure, but also more generally Western in its locatedness. This point has been well made in relation to the geographies of current analyses, and celebrations, of hybridity (Spivak, 1990; King, 1995). It applies also to some of the arguments about openness. As was pointed out above, the sudden consciousness of globalisation in the West cannot be as a result of a new 'openness' in general. What has more likely brought about the flurry of concern is the changing terms, and geography, of that openness. *Western* regions become dominated by foreign capital. The old mythical coherence of place is challenged by capital and labour from outside (not exactly a new experience, nor specific to this form of globalisation, in the majority world). It is now the West which is subject to inward investment. It is Western cities which have, in the medium term, been experiencing the arrival of people from other parts of the world. As has often been remarked, much of the work on hybridity has been stimulated by the famous 'arrival of the margins at the centre'. (This was one provocation to re-tell the history of modernity.) In that sense it is already acknowledged to be a story told from 'the first world'.

Except that, this is more of a Western story even than that account indicates. For the margins have *not* arrived at the centre. This is the view of those who were already 'in the centre' and of those from the periphery who have managed over the years to get in. Most of 'the margins' – even should they wish to migrate – have been very strictly excluded.

This is a story of globalisation which has been (as was the story of modernity) largely provoked by what is happening to the West, by the experiences of that West; it is in some measure (just as was colonial discourse) founded upon a Western anxiety. Moreover, just as in the case of modernity, this discourse of globalisation provides a legitimation of things; an imaginative geography which justifies the actions of those who promulgate it, including – and to come full circle – a particular attitude towards space and place.

My argument is that this narrative of globalisation is not spatialised. By this I do not mean simply that the picture is more geographically complex than is usually claimed: that there is significant spatial variability, or that 'the local' consistently in one way or another reasserts itself. These things are true, but they are not the argument I am making here. Indeed, Low and Barnett (2000) have accused geographers of focusing too much on this aspect of their potential contribution to the debate over globalisation. It is a focus, they argue, which reduces the discipline of geography to a concern with the local, the empirical and the a-theoretical. (I agree with the general burden of this critique. Spatialising social theory is categorically *not* reducible merely to insisting on local variation. But I remain extremely wary of any assumption of a necessary association between the terms local/empirical/a-theoretical; see Massey, 1991b.) So local variability is not what is at issue in this chapter. Rather the argument is that really 'spatialising globalisation' means recognising crucial characteristics of the spatial: its multiplicity, its openness, the fact that it is not reducible to 'a surface', its integral relation with temporality. The a-spatial view of globalisation, like the old story of modernity, obliterates the spatial into the temporal and in that very move also impoverishes the temporal (there is only one story to tell). The multiplicity of the spatial is a precondition for the temporal: and the multiplicities of the two together can be a condition for the openness of the future. Low and Barnett (2000) argue that geographers' focus on asserting 'more complex or sophisticated conceptions of space' (p. 54) (by which they mean in practice greater spatial variability) is mistaken in that we should criticise instead the *historicism* of the standard story of globalisation. My argument is that criticising the historicism of that version of the story of globalisation (its unilinearity, its teleology, etc.) precisely also entails reframing its spatiality. The reconceptualisation could (should) be of temporality and spatiality together.

But this is still one view. If space is genuinely the sphere of multiplicity, if it is a realm of multiple trajectories, then there will be multiplicities too of imaginations, theorisations, understandings, meanings. Any 'simultaneity' of stories-so-far will be a distinct simultaneity from a particular vantage point. If the repression of the spatial under modernity was bound up with the establishment of foundational universals, so the recognition of the multiplicities of the spatial both challenges that and understands universals as spatio-temporally specific positions. An adequate recognition of coevalness demands acceptance that one is being observed/theorised/evaluated in return and potentially in different terms (see, for instance, Appadurai, 2001; Slater, 1999, 2000). Recognition of radical contemporaneity has to include recognition of the existence of those limits too.

Just as the postcolonial reworking of the former story of modernity productively disrupted so much about it, so too would a genuine spatialisation of how we think about globalisation enable a very different analysis (or very different analyses) (a genuinely spatial narrative). Perhaps above all it would involve challenging that 'all-pervading denial of coevalness'. Fabian has written that it 'takes imagination and courage to picture what would happen to the West (and to anthropology) if its temporal fortress were suddenly invaded by the Time of the Other' (1983, p. 35). The same is true of so many of the ways we currently picture globalisation.

9
(contrary to popular opinion) space cannot be annihilated by time

The confusions that exist within current imaginations of the time-spaces of globalisation are, perhaps, at their most acute (and, ironically, least noticed) in the easy coexistence of the view that this is the age of the spatial with the contradictory, but equally accepted, notion that this is the age in which space will finally, in fulfilment of Marx's old prophecy, be annihilated by time.

Although clearly in conflict, these two propositions are none the less related. On the one hand, more and more 'spatial' connections, and over longer distances, are involved in the construction and understanding and impact of any place or economy or culture and of everyday life and actions. There is more 'space' in our lives, and it takes less time. On the other hand, this very speed with which 'we' can now cross space (by air, on screen, through cultural flows) would seem to imply that space doesn't matter any more; that speed-up has conquered distance. Precisely the same phenomena seem to be leading to the conclusion both that space has now won out to the detriment of any ability to appreciate temporality (the complaint of depthlessness) and that time has annihilated space.[12] Neither view is tenable as it stands.

Take, to begin with, the question of annihilation, provoked by the speed-up of global interconnections and the instantaneity of the screen. There is no doubt that recent changes on both these fronts have been enormous. Low and Barnett (2000) tell a tale of coming across, during travels in north London, a British Telecom hoarding announcing to the world that 'Geography is History'. We smile in recognition; we know what BT is getting at. (Although, and to keep the theme of ambiguity running, I have a mouse-pad which proclaims, with equal self-assurance and equal ability to seem self-evident, that 'Geography matters to all of us'. In the midst of all this contradictory confidence, it's important to keep one's nerve.) It is certainly the case that 'time' (for which read an increase in the speed of transport and communications) reduces, and indeed on occasions even annihilates, some of the effects of distance. This is what Marx was getting at. It is worth noting the irony that what is actually being reduced here is time, and what is being expanded (in the sense of the formation of social relations/interactions, including those of transport and communication) is space (as distance). This is one curiosity of the formulation. But more importantly, space is not anyway reducible to distance. Distance is a condition of multiplicity; but equally it itself would not be thinkable without multiplicity. And we might note that while cyberspace is a different kind of space (Kitchin, 1998; Dodge and Kitchin, 2001) it is most definitely internally multiple (Bingham, 1996) (and, ironically, often rendered in a language of spatial metaphor which is resolutely Cartesian). Multiplicity is fundamental. No one is proposing (I assume) that screens, or instantaneous financial transactions, or even cyberspace, are abolishing multiplicity. That would be like saying that, because a telephone call is instantaneous, the participants in it are merged into one entity. And if multiplicity is not being annihilated (which would render the whole business of transport and communication anyway entirely redundant) then neither is space. The very concept of multiplicity entails spatiality. And anyway, to complete the spectre of everything disappearing into a black hole, *how could* time annihilate space when the two are mutually implicated (see Part *Two*). So: as long as there is multiplicity there will be space.

Zygmunt Bauman has produced an elaborated version of instantaneity in his differentiation between heavy modernity (territorialising and preoccupied with size) and light: 'It all changed … with the advent of software capitalism and light modernity' (2000, p. 176). Capturing the ambiguity in the usual phrasing, he writes that 'The change in question is the new irrelevance of space, masquerading as annihilation of time. … space counts little, or does not count at all' (p. 177). 'Counting' here depends upon a notion of cost – drawing on Simmel it is proposed that things are valued to the extent of the cost of their acquisition. *Ergo*: 'If you know that you can visit a place at any time you wish', 'since all parts of space can be reached in the same time-span (that is, "no-time"), no part of space is privileged, none has special value' (p. 177). This is space as pure extension, a matter of *xy* coordinates. If space is more than (or even not) coordinates, but a product of relations, then 'visiting' is a practice of engagement, an encounter. It is in that process of establishing a relation that the 'cost' can rather be measured. (And space is made, as well as crossed, in this encounter.)

Space is more than distance. It is the sphere of openended configurations within multiplicities. Given that, the really serious question which is raised by speed-up, by 'the communications revolution' and by cyberspace, is not whether space will be annihilated but what kinds of multiplicities (patternings of uniqueness) and relations will be co-constructed with these new kinds of spatial configurations.

One aspect of this radical reordering of the co-constitution of space and difference is already much discussed. Among the many other currently popular aphorisms about space and time are the propositions (i) that there is no longer any distinguishing between near and far and (ii) that the margins have invaded the centre.

There is, as has been seen, a way of understanding the rise and fall of modernity in terms of a founding moment in which difference from 'the rest of the world' was established by the West either through temporal convening or through territorialisation. The collapse of (or challenge to) that sensibility was provoked by the impossibility of maintaining the story in the face of the breakdown of the geography it purported to describe: the margins arrived at the centre, those who had been far away were now very evidently near (in both space and time).

There is much to be said for this interpretation: it has run as a thread through much of Part *Three*. Indeed, I would interpret it as modernity's way of taming the disruptiveness of the spatial, and subsequently its inability to maintain that feeling of control over things (the failure of its political cosmology) when 'real geographical space' (which had always in fact failed to conform) now failed to conform to such an extent that the ordering framework could no longer hold.

This is, then, a good way of capturing some important aspects of the constitution of modernity and whatever it is that we are experiencing now. It must, however, be treated carefully. To begin with, who is this 'we'? Countries on the end of colonialism, invasion, the long history of European multinational economic exploitation, are not now for the first time experiencing the arrival of the previously distant. The collapse of near and far has long been a fact for places *outside* the West – indeed it is intrinsic to the establishment, through 'discovery', imperialism and colonialism, of modernity itself. Moctezuma would attest to that. Once again the Western roots of the dominant sensibility are evident. The tale of the arrival of the margins at the centre needs similar interrogation. Here, not only is the shift in sensibility, the breakdown of the old ordering mechanisms, quite explicitly located in the West, but the empirical basis is itself questionable. The margins have not arrived in the centre.

Among the more complex versions of this story one strategy has been to develop an argument concerning the relation between distance and otherness. Rob Shields (1992), while more healthily sceptical than many about the passage from one 'space-time regime' to another, argues that we are witness to a significant shift in one aspect of social spatialisation. His argument is that, through the institution of its particular global geography, there developed within modernity a strong association between presence/absence on the one hand and inclusion/exclusion on the other. This has now been upset through changes in which 'the interpenetration of cultures and the increased presence of distant "others" in everyday life in the developed countries of the West are perhaps the key driving forces' (p. 193). A 'postmodern spatialisation' comes on to the agenda.

Now Shields is absolutely scrupulous in his insistence on the recognition of the spatio-temporal specificity both of the socio-economic changes and of the shifts in dominant sensibilities. Indeed, he strongly criticises others for not being so: 'Giddens (in what is by now a tradition of ethnocentric error amongst Western social scientists) installs historically specific, modernist forms and self-interpretations as universals' (p. 192; the reference is to Giddens, 1984). His own argument, however, raises questions of another sort. His argument is that, under modernity, and integral to its very establishment/nature, 'inclusion and exclusion are meshed with the terms of proximity and remoteness, presence and absence' (p. 192) and that with postmodern spatialisation 'The distances that once separated all the categories of "otherness" from the local sphere of "our" everyday life appear to have collapsed or are at least undergoing important changes' (p. 194).[13] But not all the 'others' whose existence and difference were so vital to the establishment of the modern sensibility were located in distant regions of the planet. There were also 'others' within: not least, though also not only, 'women' and 'nature'. McClintock (1995) has explored the interweaving of race, gender and class in the establishment of British imperialism. Haraway (1991) has pointed to the significance of the excluded figures of the feminine, the animal and the mechanical. Even within modernity there were many modes of establishing otherness (exclusion), not all of them dependent on distance.

The argument here is simply that what is, or should be, at issue in accounts of modernity and of globalisation (and indeed in the construction/conceptualisation of space in general) is not a kind of denuded spatial form in itself (distance; the degree of openness; the numbers of interconnections; proximity, etc. etc.), but the relational content of that spatial form and in particular the nature of the embedded power-relations. There is no mechanical correlation between distance and difference. Both the othering of the rest of the world and the othering of femininity within the establishment of the classic figure of modernity employed the manipulation of spatiality as a powerful tool, but the kinds of power which were involved, and the ways in which these were enforced through the configuration of the spatial, were in each case quite different (see Massey, 1996a). Spatiality was important in both cases; but space is more than distance. Location, confinement, symbolism … played their roles too. What is at issue is the articulation of forms of power within spatial configurations.

Indeed, it may be through the establishment of new power-invested spatial configurations, rather than simply through the conquering of distance by speed-up, that the challenging of certain characteristics of spatiality is potentially on the agenda. One of the things which 'cyberspace' most famously allows is instantaneous contact at a distance. This is, moreover, both networked and selective. The connections can be multiple and you can choose with whom you are in contact (the latter is, of course, not entirely the case, a fact which ironically – see below – may be a saving grace). Communities, in the sense of networks of communication of common interest, of similarity along selected dimensions, can easily be established at a distance; non-contiguous time-spaces of commonality. But there are forebodings too. Kevin Robins (1997) has written persuasively of some of these. While the protagonists of what he calls 'the new politics of optimism' – Bill Gates (1995), Nicholas Negroponte (1995), William Mitchell (1995) – talk of the possibility of electronically overcoming social division, Robins is more cautious. What this politics of optimism involves is an assumption, not only of space as merely distance, but also of it as always *a burden*. It is persistently characterised, in these discourses, as a constraint. (The constraint of distance, rather than, perhaps, the pleasure of movement or travel.) Says Negroponte, 'the post-information age will remove the limitations of geography' (1995, p. 165, cited in Robins, 1997, p. 197). As Robins puts it:

> The politics of optimism wants to be rid of the burden of geography (and along with it the baggage of history), for it considers geographical determination and situation to have been fundamental sources of frustration and limitation in human and social life. (p. 198)

There has, posits Robins, been 'a longstanding desire for transcendence' of this earthboundness; of the 'constraints of space and place' (p. 198), and he argues for caution in terms of the notions of communication and community (and the idealised versions, both frictionless and nostalgic, imagined by the digital optimists) and in terms also of the significance of materiality (as opposed to virtuality).

One aspect of this argument is that as our long-distance communications increase so the significance may decrease of those who live next door. ('We will socialise in digital neighbourhoods in which physical space will be irrelevant' – Negroponte, 1995, p. 7, cited in Robins, p. 197.) And that precisely would be to undermine one of the truly productive characteristics of material spatiality – its potential for the happenstance juxtaposition of previously unrelated trajectories, that business of walking round a corner and bumping into alterity, of having (somehow, and well or badly) to get on with neighbours who have got 'here' (this block of flats, this neighbourhood, this country – this meeting-up) by different routes from you; your being here together is, in that sense, quite uncoordinated. This is an aspect of the productiveness of spatiality which may enable 'something new' to happen. It also poses questions in the sphere of the social. It is against this uninvited juxtaposition that the battles for 'the purification of space' are waged, whether through the employment of security guards around the gated communities of the privileged, through controls over international migration or – for these battles are not always about the powerful excluding the weak – through attempts to preserve some space of their own by groups which are socially marginalised. We may support one side or the other – the issue is one of spatialised power not of abstract form – but what is important is that contact is involved and some form of social negotiation. What cyberspace, on some readings, could potentially enable is a kind of disembedding into non-contiguous communities of people-like-us which evade all those challenges thrown up by what material spatiality always presents you with – the accidental, unchosen (different) neighbour. Viewing space as a matter only of distance, and then in that guise only negatively as a constraint, lies behind what may be a tendency to try to escape one of its most productive/disruptive elements – one's different neighbour. Staple (1993) has written of a 'new tribalism'. 'Conquering' distance in no way annihilates space, but it does raise new issues around the configuration of multiplicity and difference.

This is absolutely not a sentimental plea for the joys of mixed localities, or for the simple locatedness of place. (Indeed an alternative approach to place is proposed in the next chapter. And these arguments about closeness across physical distance also have the significant political potential, from a geographical point of view, of disrupting that old assumption that one's priorities in terms both of affect and of responsibility begin close in – your family, your neighbourhood – and then, with decreasing resonance, spread outwards in concentric circles.) Rather, what is being signalled here is a concern about a potential new dimension of gatedness. If the previously far really is getting too near for your comfort, if in your view the margins really are too much invading the centre, then in addition to wielding the mechanisms of market forces and discrimination in reorganising your location and choosing your neighbours you can now extricate yourself even more, by living at least some of your life in another purified space, on the Net.

Except … Except that 'space' won't allow you to do it. Space can never be definitively purified. If space is the sphere of multiplicity, the product of social relations, and those relations are real material practices, and always ongoing, then space can never be closed, there will always be loose ends, always relations with the beyond, always potential elements of chance. Indeed, again, this set of characterisations of the current era is rivalled by its opposite – tales of hybridity, mixity, of hackers, invasions, viruses and flux. All of them utterly ambiguous, of course; but that is the point – neither hermetic closure nor a world composed only of flow (no stabilisations, no boundaries of any sort) is possible. While the end of cities through technology-led dispersal is confidently predicted by cyberfuturists, cities are growing as never before (Graham, 1998). Mobility and fixity, flow and settledness; they presuppose each other. As Saskia Sassen (2001) points out, the global city itself, with its enormous capacity for generating and controlling flows, is built upon vast emplaced resources. The impetus to motion and mobility, for a space of flows, can only be achieved through the construction of (temporary, provisional) stabilisations. There is only ever, always, a negotiation (and a responsibility to negotiate) between conflicting tendencies. A restructuring of the geography of that simultaneity of stories-so-far. This is not the annihilation of space; but it is a radical reorganisation of the challenges that spatiality poses.

And anyway the tales of cyberspace are belied by its own, very material, necessities. The devaluation of space and place which runs through this literature is one aspect of a general shift by which 'information' has been conceptualised as disembedded from materiality, one implication of which has been 'a systematic devaluation of materiality and embodiment' (Hayles, 1999, p. 48). For all that so many of the tales of the effects of cyberspace revolve around its ability to render space insignificant, in the context of its own material production and operation (on the ground, as it were) space is of fundamental importance. The producers of cyberspace actually know very well that space is more than distance, and that it matters crucially. The science parks and similar enclosures of high-technology production are knowingly created enclaves: set apart from the messy world, devoted to a single activity (the production/elaboration, and glorification of high technology), purified quite rigorously although never entirely successfully of 'non-conforming' uses (those which would interfere, not just with process, but with image), acutely aware of location, and often quite elaborately guarded. And not only are they regulated in a physical sense, they are also very deliberately about meaning: the interaction between the status of the scientists and the locational cachet of the place upholds the authority of social status, of the place and of the science itself (Massey, 1995b; Massey et al., 1992). This is space as multiplicity and hence of heterogeneity and uniqueness. The contrast between the supposed effect of cyberspace and the dynamics of its own production – between, that is, the overcoming of space on the one hand and a supremely nuanced use and making of it on the other – precisely highlights the difference between space understood only as distance and space in a richer meaning. Whatever is happening to the former the latter is very far from being annihilated. And this fact that the virtuality of cyberspace has its roots very firmly in the ground highlights something else as well: that the world of physical space and the world of electronically mediated connection do not exist as somehow two separate layers, one (in what is I suspect a common mind's eye imagination) floating ethereally somewhere above the materiality of the other. As Rob Kitchin (1998) has argued: 'cyberspatial connections and bandwidth … are unequally distributed [spatially]'; 'information is only as useful as the locale within which the body resides'; and 'cyberspace depends on real-world spatial fixity – the points of access, the physicality and materiality of wires' (p. 387). Or again, for Stephen Graham 'power to function economically and link socially increasingly relies on constructed, place-based, material spaces intimately woven into complex telematics infrastructures linking them to other places and spaces' (1998, p. 174; see also Pratt, 2000). Just as the groundedness of virtuality ties it to a specificity of location so too spaces and places are altered in their physicality and in their meaning through their embeddedness in networks of communication. The 'virtual' world depends on and further configures the multiplicities of physical space. This has ever been so; the new media in that sense are not new, but they do refigure (or have the potential to refigure) how those networks will operate.

Graham (1998) has usefully distinguished between three modes of conceptualising the relationship between information technology, space and place. First, there is the mode, which we have considered above, that he characterises as 'substitution and transcendance: technological determinism, generalized interactivity and the end of geography', and which he roundly criticises for its naive technological determinism. Second is the mode of 'co-evolution: the parallel social production of geographical space and electronic space' which, rejecting technological determinism, argues that electronic and territorial spaces are necessarily produced together. Third, there is the mode of 'recombination' which involves the mutual constitution of technology and the social sphere (see, for instance, Callon, 1986; Haraway, 1991; Latour, 1993; Pratt, 2000). It is within this third mode of mutual constitution, he argues, that we can most aptly understand the continual remaking of space.

Moreover, and as the authors of the 'recombination' approach have long argued, 'mutual constitution' is not between the human and the technological alone, but with (what we choose to call) 'nature' too. If the mantras around new technology have evoked an infinite instantaneity of dematerialised mobility those around nature have proposed the opposite. As Clark (2002) points out, while we recognise the mobility in culture and society there is a tendency to be unnerved by the mobility of nonhuman life. Cheah (1998) makes a related point about 'hybridity theorists' (p. 308). We worry about the 'unnatural' mixings we are producing in the 'natural' world: 'Social and cultural theorists are taking global ecological despoliation as evidence of a general de-naturalization that now encompasses the biophysical world in its entirety' (Clark, 2002, p. 103). While this recognises co-constitution it works also with a background assumption that the 'natural' world if left to itself would somehow, still, really, be organised through that modernist territorial spatiality, settled into its coherent regions in rooted indigeneity.

> But why is it exactly, we might wonder, that there is so much political purchase to be had from the idea of nature's undoing at the hands of culture, and so little currency in considering the things life achieves on its own account? … And why is it that after all the vexing of the nature/culture binary, we are still so much more comfortable tracking the impact of globalization *on* the biophysical world than we are with any consideration of a biological or geological contribution *to* the global contours we now confront? (2002, p. 104; my emphases)

And 'though it may be true that the ecologically aware, while acting locally have tried to "think globally", this gesture has tended to involve a planet-scale projection of qualities of homeliness and rootedness' (p. 105). Clark diagnoses this as a perspective from the cities of Europe and the USA: 'both its constitutive strands – the environmentalist belief in a nature which "stays put" and the cosmopolitan celebration of culture free of groundedness and material responsibilities – can be seen as derivatives of the same metropolitan detachment from the daily dynamics of bio-materiality' (p. 117). (He offers the experience of the colonial periphery as one alternative.)

Understanding nature as essentially 'staying put' is a manoeuvre that hints at a desire for a foundation; a stable bottom to it all; a firm ground on which the global mobilities of technology and culture can play. The global flows of the planet, organic and inorganic, prohibit any ultimate refuge of this kind. Clark takes 'the now routine insistence on the porosity of the nature/culture binary at its word' and proposes that 'the notion of "globalization from below" might have new connotations if it can be shown that there is no final cut-off point to this "below", no guard-rail to keep us to the realm of the already humanized' (p. 105). And once that has been taken into account, somehow all the excitement about so-called instantaneity and speed-up dies away and they are reduced to their more proper position within a planet that has ever been a global mobility.

10
elements for alternatives

Whether or not it is the case that these are peculiarly spatial times, the conceptualisation of space itself is, crucially but usually implicitly, a stake in emerging confrontations. Richard Peet (2001), in his thoughtful review of MacEwan's *Neoliberalism or democracy?* (1999), has argued that it is necessary to deepen still further the critique of neoliberalism and the political project in which it is embedded. The argument here is that attention to the implicit play of contesting understandings of space could be integral to this project. It could be central to his suggestion that we need 'to reveal neoliberalism as a discourse structured, eventually, by multinational corporations … and to read neoliberal hegemony geographically' (p. 340). Neoliberal globalisation as material practice and as hegemonic discourse is yet another in a long line of attempts to tame the spatial. Nor is this only a matter of critique. Attention to implicit conceptualisations of space is crucial also in practices of resistance and of building alternatives.

It has been argued here that many current discourses around globalisation evade the full challenge of space. Convening spatial heterogeneity into temporal sequence deflects the challenge of radical contemporaneity and dulls the appreciation of difference. Equating space with depthless instantaneity deprives it of any dynamic. Envisioning space as always-already territorialised, just as much as envisioning it as purely a sphere of flows, misunderstands the ever-changing ways in which flows and territories are conditions of each other. It is the practices and relations which construct them both that demand address. In contrast, and building on the arguments of Part *Two*, what have been stressed here are other characteristics. First, space as the sphere of heterogeneity. Position, location, is the minimum order of differentiation of elements in the multiplicity that is co-formed with space. It is thereby also the condition for a more radical heterogeneity. Grossberg has written of the need for space to become a philosophical project and argued that, within such a project, 'spatializing the real' would mean conceptualising 'the real as the production of the singularity of the other' (1996, p. 179). Second, space as the sphere of relations, negotiations, practices of engagement, power in all its forms (Allen, 2003). In this context, space is the dimension which poses the question of the social, and thus of the political (while 'actual' spaces are produced *through* the social and the political). And third, space as the sphere of coevalness, of radical contemporaneity.

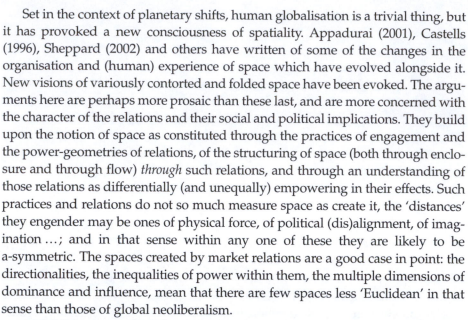

Set in the context of planetary shifts, human globalisation is a trivial thing, but it has provoked a new consciousness of spatiality. Appadurai (2001), Castells (1996), Sheppard (2002) and others have written of some of the changes in the organisation and (human) experience of space which have evolved alongside it. New visions of variously contorted and folded space have been evoked. The arguments here are perhaps more prosaic than these last, and are more concerned with the character of the relations and their social and political implications. They build upon the notion of space as constituted through the practices of engagement and the power-geometries of relations, of the structuring of space (both through enclosure and through flow) *through* such relations, and through an understanding of those relations as differentially (and unequally) empowering in their effects. Such practices and relations do not so much measure space as create it, the 'distances' they engender may be ones of physical force, of political (dis)alignment, of imagination …; and in that sense within any one of these they are likely to be a-symmetric. The spaces created by market relations are a good case in point: the directionalities, the inequalities of power within them, the multiple dimensions of dominance and influence, mean that there are few spaces less 'Euclidean' in that sense than those of global neoliberalism.

And this is a space, too, that is forever incomplete and in production. Its openness (ironically, the very difficulty of its representation – its 'ungraspability' in Jameson's terms) is the other aspect of its challenge. The openended interweaving of a multiplicity of trajectories (themselves thereby in transformation), the concomitant fractures, ruptures and structural divides, are what makes it in the end so unamenable to a single totalising project. Castells' cultural and spatial discontinuities, his populations and places of 'structural irrelevance', Appadurai's disjunctures … even the new hybridities formed at points of intersection and juxtaposition are just as much a product of the dissonances, absences and ruptures within the process of globalisation as of any simple increase in the building of interconnections. If, then, we were to draw a map of the new globalisation (even quite an ordinary map of flows, say) it would not show a totally interconnected system: there would be both long-standing absences and the systematic production of new disconnections. This is not meant to imply the existence of autonomous islands (*not* a re-evocation of a billiard-ball geography) – only the geography of globalisation is at issue here; there will be other connections. Such disjunctive moments will take on different names in different vocabularies, and will have distinct inflections (a clash of differences which remain untotalisable; the undetermined futurity of a conjuncture), but they share an openness in which there is still room for politics.

Most importantly perhaps this is to take up Fabian's challenge in the face of a hegemonic imagination of globalisation in which, to transpose to this context Fabian's own words, 'the all-pervading denial of coevalness … ultimately is expressive of a cosmological myth of frightening magnitude and persistency' (1983, p. 35).[14]

Even such a hasty sketch raises questions for a politics around neoliberal globalisation. I want to focus here on just three elements of this: relationality, implication and specificity. Most obviously, as already argued, a bi-polarisation of a space of free movement on the one hand and a space of enclosed territories on the other is not only a contradiction which it is important to highlight in the current conservative/neoliberal constellation, it may also be dangerous ground for the construction of opposition and/or alternatives. On the one hand, this is so for the old reason of spatial fetishism – abstract spatial form in itself can guarantee nothing about the social, political or ethical content of the relations which construct that form. What is always at issue is the *content*, not the spatial form, of the *relations through which* space is constructed. But the issue is also more serious than this. There is an overwhelming tendency both in academic and political literature, and other forms of discourse, and in political practice to imagine the local as the product of the global but to neglect the counterpoint to this: the local construction of the global. 'Local places' in a general sense, whether they be nation-states or cities or small localities, are characteristically understood as *produced through* globalisation. There are problems on both sides of this counterposition. On the one hand, it is to understand the global, implicitly, as always emanating from somewhere else. It is therefore unlocated; nowhere. This has direct parallels with that imagination of information as disembedded and disembodied (Hayles, 1999). On the other hand, local places, in this understanding of globalisation, have no agency. As Arturo Escobar characterises the classic mantra: 'the global is associated with space, capital, history and agency while the local, conversely, is linked to place, labor, and tradition – as well as with women, minorities, the poor and, one might add, local cultures' (2001, pp. 155–6). Place, in other words, is figured as inevitably the *victim* of globalisation.[15]

There has, in recent years, been something of a fightback on this front, and an assertion of the potential agency, within the context of neoliberal globalisation, of 'local place' (Dirlik, 1998; Escobar, 2001; Gibson-Graham, 2002; Harcourt, 2002). Even these important statements have, none the less, remained within a discourse of 'the defence of place', of a political defence of the local against the global.

However, taking seriously the relational construction of space points to a more variegated politics. For in a relational understanding of neoliberal globalisation 'places' are criss-crossings in the wider power-geometries that constitute both themselves and 'the global'. On this view local places are not simply always the victims of the global; nor are they always politically defensible redoubts *against* the global. Understanding space as the constant open production of the topologies of power points to the fact that different 'places' will stand in contrasting relations to the global. They are differentially located within the wider power-geometries. Mali and Chad, most certainly, may be understood as occupying positions of relative powerlessness. But London, or the USA, or the UK? These are the places in and through which globalisation is *produced*: the moments through which the global is constituted, invented, coordinated. They are 'agents' *in* globalisation. This is not to say that 'whole places' are somehow actors (see later) but it is to urge a politics which takes account of, and addresses, the *local* production of the neoliberal capitalist *global*.

There are a number of immediate implications. To begin with, this fact of the inevitably local production of the global means that there is potentially some purchase through 'local' politics on wider global mechanisms. Not merely defending the local against the global, but seeking to alter the very mechanisms of the global itself. It raises the question of local 'responsibility' for the global – which will be addressed in Part *Five*. Different places occupy distinct positions within the wider power-geometries of the global. In consequence, both the possibilities for intervention in (the degree of purchase upon), and the nature of the potential political relationship to (including the degree and nature of responsibility for) these wider constitutive relations, will also vary. It is no accident that much of the literature concerning the defence of place has come from, or been about, either the South or, for instance, deindustrialising places in the North. From such a perspective, capitalist globalisation does indeed seem to arrive as a threatening external force. But in other places it may well be that a particular construction of place is *not* politically defensible as part of a politics against neoliberal globalisation – and this is not because of the impracticality of such a strategy but because the construction of that place, the webs of power-relations through which it is constructed, and the way its resources are mobilised, are precisely what must be challenged.

This, then, would be a local politics that took seriously the relational construction of space and place, and as such would be highly differentiated through the vastly unequal articulation of those relations. The local relation to the global will vary and in consequence so will the coordinates of any potential local politics of challenging globalisation. Indeed, to argue for the defence of place in an undifferentiated manner is in fact to maintain that association of the local with the good and the vulnerable to which both Escobar and Gibson-Graham quite rightly object.

What, in the end, is of concern here is a persistent tendency to exonerate the local. Bruce Robbins (1999), musing upon forms of 'American' nationalism which have achieved respectability, argues that

> One distinctive feature is that capitalism is attacked only or primarily when it can be identified with the global. Capitalism is treated as if it came from somewhere else, as if Americans derived no benefit from it – as if … American society and American nationalism were among its pitiable victims. … By refusing to acknowledge that these warm insides are heated and provisioned by that cold outside, these avowedly anticapitalist critics allow the consequences of capitalism to disappear from the national sense of responsibility'. (p. 154)

Exactly the same argument could be made about many another place constructed as a node of power within global geometries. What is problematical politically is that a persistent defence of the local, *qua* the local, without regard to the constitutive social relations, can lead to a lack of address to the *constitution* of the local itself.

One important thread in this argument is that conceptualising space in terms of practices and relations raises the question of *implication*. The local is implicated in the production of the global. Moreover, taking this seriously fundamentally challenges some of the most persistent metaphorical 'geographies of resistance'. The discussion of de Certeau's conceptualisation of space and time in Part *Two* has already raised this issue. There the formulation was in terms of the little tactics of the street in some way resisting 'the proper place' of power. 'Power' and 'resistance' in the very imagination of their spatial separation in this way are also constituted separately. There is no opportunity, in this structure, to examine the relations between them (see also on this Sharp et al., 2000). In like manner, the imaginations of 'resistance' in terms of a spatiality of 'margins', or of 'interstices', block off more serious political engagement. They are all, anyway, forms of spatial fetishism, assuming a politics from a geography. They play out a romance of detachment which refuses to recognise any implication in this 'power', or to take responsibility for it. And by doing this, they lose a possible point of purchase for an effective politics.

And finally, such an understanding of the nature of globalised space points to a politics of *specificity*. As was argued above, a local–global politics would be structured differently from place to place. Moreover, this recognition of specificity is necessary too even in the face of global institutions. This argument runs quite against the current grain of thinking. Thus, the World Trade Organisation operates through the implementation of rules (the rules of free trade, etc.) which claim fairness on grounds of their universal application. Yet, evidently, the application of equal, abstract rules, in a world of endless specificity, not to mention gross inequality, is not in fact 'fair'. That kind of apparent evenhandedness will never produce the egalitarian outcomes that are claimed for them. It follows that the argument that the rules of 'free trade' should be applied more fairly (that the EU should abandon the quotas on textiles; the USA the subsidies of cotton production, etc.) is right (because at the moment the rules are bent in favour of the powerful), but it is not enough. Arguments against free trade are similarly inadequate – protectionism may be justifiable or not, depending on the power-relations constructing each specific situation ('protectionism' is another of those words, like globalisation, which has been captured by the political right). In order to respond to specificity, however, one needs (ever-provisional) agreement about aims, and that requires global fora of a very different nature. They would need to be fora which could debate purposes, and argue over the *form* of globalisation in relation to those purposes (Massey, 2000a, 2000b), and respond to individual instances in a situated way within those wider premises. The objection to such a suggestion would undoubtedly be that it would lead to endless debate and disagreement. And it undoubtedly would. But endless debate and disagreement are precisely the stuff of politics and democracy. (The effect of the application of 'rules' is that, as with the assertion of the inevitability of globalisation, it takes politics out of the debate. It treats the process of globalisation as a technical matter.) Understanding globalisation through the specifics of the geometries of power enforces its politicisation, beyond the terms of for it or against it and around the terms of what it's for and what form it's going to take.

Part *Four*
Reorientations

Whether it be poring over maps, taking the train for a weekend back home, picking up on the latest intellectual currents, or maybe walking the hills … we engage our implicit conceptualisations of space in countless ways. They are a crucial element in our ordering of the world, positioning ourselves, and others human and nonhuman, in relation to ourselves. This Part explores a mixture of these things: routine material practices, certain common tropes and attitudes, and one or two particular texts. What space gives us is simultaneous heterogeneity; it holds out the possibility of surprise; it is the condition of the social in the widest sense, and the delight and the challenge of that.

11
slices through space

Falling through the map

I love maps – they are one of the reasons I became 'a geographer'. They carry you away; they set you dreaming. Yet it may well be none the less that our usual notion of maps has helped to pacify, to take the life out of, how most of us most commonly think about space. Maybe our current, 'normal' Western maps have been one more element in that long effort at the taming of the spatial.

Faced with a need to know (just where exactly is Uzbekistan? What is the layout of this town? How *do* I get from here to Ardwick?) you reach for the map and lay it out upon the table. Here is 'space' as a flat surface, a continuous surface. Space as the completed product. As a coherent closed system. Here space is completely and instantaneously interconnected; space you can walk across. The map works in the manner of the synchronies of the structuralists. It tells of an order in things. With the map we can locate ourselves and find our way. And we know where others are as well. So yes, this map can set me dreaming, let my imagination run. But it also offers me order; lets me get a handle on the world.

Are maps an archetype of representation? We 'map things out' to get a feeling for their structure, we call for 'cognitive maps',[1] 'we' (or so I read in reliable sources) are currently 'mapping' DNA. Maps as a presentation of an essential structure. The ordering representation.

But our notion of the root meaning of 'map', the term map in its most common current Western usage, has to do with geography and hence with space. So all the conflations get run together, are conflated in their turn. Maps are about space; they are forms of representation, indeed iconic forms; representation is understood as spatialisation. But a map of a geography is no more that geography – or that space – than a painting of a pipe is a pipe.

Obviously maps are 'representations'. And they are so in the sophisticated, creative, sense in which we have learned to mean that word. Obviously, and inevitably too, they are selective (as is any form of re-presentation). This is Borges' old point. Moreover, through their codes and conventions and their taxonomic and ordering procedures, maps operate as a 'technology of power'

(Harley, 1988, 1992). But it is not those things that are important to me here. It is not even – as we lay the map (the country we shall visit, the town, the region to be conquered) out on the table before us – the much-maligned notion of 'the view from above'. Not all views from above are problematical – they are just another way of looking at the world (see the disagreement with de Certeau in Chapter 3). The problem only comes if you fall into thinking that that vertical distance lends you truth. The dominant form of mapping, though, does position the observer, themselves unobserved, outside and above the object of the gaze. None the less, what worries me here is another and less-recognised aspect of this technology of power: that maps (current Western-type maps) give the impression that space is a surface – that it is the sphere of a completed horizontality.

But what if – recalling the arguments of Part *Two* – the assumption is abandoned that space and time are mutually excluding opposites? What if space is the sphere not of a discrete multiplicity of inert *things*, even one which is thoroughly interrelated? What if, instead, it presents us with a heterogeneity of practices and *processes*? Then it will be not an already-interconnected whole but an ongoing product of interconnections and not. Then it will be always unfinished and open. This arena of space is not firm ground on which to stand. In no way is it a surface.

This is space as the sphere of a dynamic simultaneity, constantly disconnected by new arrivals, constantly waiting to be determined (and therefore always undetermined) by the construction of new relations. It is always being made and always therefore, in a sense, unfinished (except that 'finishing' is not on the agenda). If you really were to take a slice through time it would be full of holes, of disconnections, of tentative half-formed first encounters. 'Everything is connected to everything else' can be a salutary political reminder that whatever we do has wider implications than perhaps we commonly recognise. But it is unhelpful if it leads to a vision of an always already constituted holism. The 'always' is rather that there are always connections *yet to be* made, juxtapositions yet to flower into interaction, or not, potential links which may never be established. Loose ends and ongoing stories. 'Space', then, can never be that completed simultaneity in which all interconnections have been established, in which everywhere is already (and at that moment unchangingly) linked to everywhere else.

Loose ends and ongoing stories are real challenges to cartography. Maps vary of course. On both sides of the Atlantic before the Columbian encounter maps integrated time and space. They told stories. While presenting a kind of picture of the world 'at one moment' (supposedly) they also told the story of its origins. *Mappae mundi* advertised the world as having Christian routes, and produced a cartography which told the Christian story. On the other side of the Atlantic, in what was to become the Americas, Toltecs, Mixteca-Puebla and other groups designed cartographies which accounted for the origins of their cosmos. In the Codex Xolotl, mentioned in Part *One*, 'Events are choreographed' (Harley, 1990, p. 101). These are maps which recount histories, which integrate

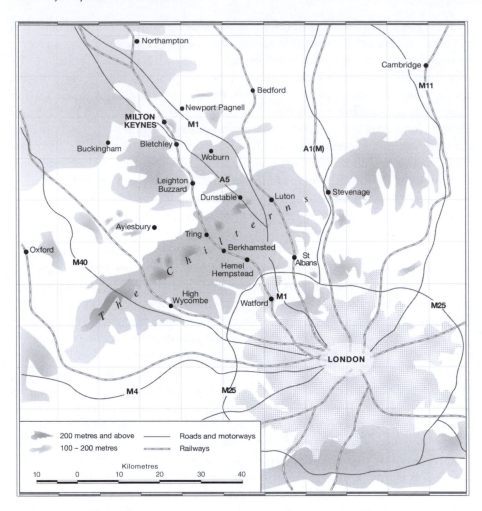

figure 11.1   *Ceci n'est pas l'espace*

time and space. There is an irony here. This turning of a migration into a line on a map, the line of footsteps on the Codex Xolotl, is one of the many routes by which representation has come to be called spatialisation. A movement is turned into a static line. Chapters 2 and 3 explored this, though it is nice to add here that part of de Certeau's argument, concerning his decision not to use the term trajectory, is neatly countered by the Codex map – the directionality of the footsteps makes it clear that there is no reversibility here: you can't go back in space-time. However these maps recall a further point from Part *Two*. These are 'representations' of space *and* time. It is not the spatial which is fixing the temporal but the map (the representation) which is stabilising time-space.

And stabilisation, or at least getting (being given) one's bearings in a universe, and in many cases making a claim on it, was what these maps were all about. They were the hegemonic cognitive mappings of five hundred years ago. They were attempts to grasp, to invent, a vision of the whole; to tame confusion and complexity.

Some mappings, on the other hand, work to do the opposite, to disrupt the sense of coherence and of totality. Situationist cartographies, while still attempting to picture the universe, map that universe as one which is not a single order. On the one hand, situationist cartographies sought to *dis*orient, to defamiliarise, to provoke a view from an unaccustomed angle. On the other hand, and more significant to the argument here, they sought to expose the incoherences and fragmentations of the spatial itself (in their case primarily the space of the city). This is the opposite of the synchronies of the structuralists: a representation of geographical space, not an a-spatial conceptual structure. Here there is exposure rather than occlusion of the disruptions inherent in the spatial. Here the spatial is an arena of possibility. Such a cartography attempts what Levin has called a mimesis of incoherence (Levin, 1989, cited in Pinder, 1994). It is a map (and a space) which leaves openings for something new.

So, most certainly, space is not a map and a map is not space, but even maps do not have to pretend to entail coherent synchronies.

More recently there have been other experiments. 'The figure of cartography recurs in contemporary cultural theory', writes Elizabeth Ferrier (1990, p. 35); '… [m]apping seems to be crucial to postmodernity'. The figure of the map has been taken up in some postcolonial and feminist literature as a form that can on the one hand stand for past rigidities but that can also, on the other hand, be reworked from within (Huggan, 1989). In these projects, maps can be both deconstructed and then reconstructed in a form which challenges the claims to singularity, stability and closure which characterise our usual notion of (and indeed in most cases the intentions of) cartographic representation.

Here, the Derridean opening up of representation is brought to bear on the classic form of the Western, modern map. The production of such maps is an 'exemplary structuralist activity', writes Huggan (1989, p. 119). They are conceptual and a-temporal – but ironically, given that these are maps, they are not spatial – structures. Huggan draws on Derrida's notion of contradictory coherence to argue that maps of this sort necessarily trace 'back to a "point of presence" whose stability cannot be guaranteed' (p. 119). The 'synchronic essentialism' of such maps may thus be opened up, and thereby the closure to which they – and their makers – aspire may be challenged from within. It is a challenge which aims to unsettle 'the classic Western map' in a number of ways. On the one hand, it disputes the internal coherence, the singular uniformity, to which the classic map lays claim – it points to the 'blind spots', the 'forgetfulness of antecedent spatial configurations' (Rabasa, 1993), the 'discrepancies and approximations' (Huggan, 1989) which cannot be obliterated. In other words, the hints of multiplicity. On the other hand, the deconstructive challenge recognises a necessary provisionality and transitoriness which undermines the claims to fixity, to pinning things down, which characterise the classic Western modern map. What is going on here then – in these feminist and postcolonial reimaginings of the possibilities of cartography – is a pushing further of the critique of maps as 'technologies of power' to lever open our understanding of the form of the map itself.

And yet … 'blind spots', the 'forgetfulness of antecedent spatial configurations' and, from Spivak, the coloniser's 'necessary yet contradictory assumptions of an uninscribed earth' (1985, p. 133) all draw, in the postcolonial context, on the notion of the colonial text as writing over a thereby obliterated other. They figure multiplicity through the form of a palimpsest. This can capture the strategy of domination as well as hinting at the possibility of disruption. Thus Rabasa: 'the image of the palimpsest becomes an illuminative metaphor for understanding geography as a series of erasures and overwritings that have transformed the world. The imperfect erasures are, in turn, a source of hope for the reconstitution or reinvention of the world from native and non-Eurocentric points of view' (1993, p. 181). It is this imperfect erasure which can be 'perhaps also a means of delineating a series of blind spots from which counter-discourses to Eurocentrism may take form' (p. 183). Yes; but while this deconstructive strategy may enable critique of colonial discourses and a pointing towards other voices, other stories for the moment suppressed, its imagery is not one which easily provides resources for bringing those voices to life. This is one of the reservations of Rajchman (1998) in his retrospective critique of collage and superposition (Part *Two*, Chapter 4). For while being critical of the layer of apparent coherence laid over alternative voices by the dominant power (in postcolonial terms the power of Europe; in more general terms the power of the maker of maps of this form), it continues to imagine the heterogeneous multiplicity in terms of layers. Yet 'layers' (as in 'the accretion of layers') would seem rather to refer to the history of a space than to its radical contemporaneity. Coevalness may be pointed to, but it is not established, through the metaphor of palimpsest. Palimpsest is too archaeological. In this story, the things that are missing (erased) from the map are somehow always things from 'before'. The gaps in representation (the erasures, the blind spots) are not the same as the discontinuities of the multiplicity in contemporaneous space; the latter are the mark of the coexistence of the coeval. Deconstruction in this guise seems hampered by its primary focus on 'text', however broadly imagined. To picture this argument through the figure of the palimpsest is to stay within the imagination of surfaces – it fails to bring alive the trajectories which co-form this space. Thus Rabasa writes of 'the strata of palimpsests underlying cartography' (p. 182). But this is to imagine the space being mapped – which is a space as one simultaneity – as the product of superimposed horizontal structures rather than full contemporaneous coexistence and becoming.

Situationist cartographies, more recent deconstructions, attempts to think in rhizomatic terms, all are wrestling to open up the order of the map. Deleuze and Guattari, in combat against the pretensions both to representation and to self-enclosure, distinguish between a tracing (an attempt at both) and 'the map' which 'is entirely oriented toward an experimentation in contact with the real. … It is itself a part of the rhizome' (1987, p. 12). But within the dominant understanding of the space of the 'ordinary' map in the West today the assumption is precisely that there is no room for surprises. Just as when space is understood *as* (closed/stable) representation (the 'spatialization' through which 'surprises are averted', de Certeau, 1984, p. 89), so in this representation *of* space you never lose your way, are never surprised by an encounter with something unexpected, never face the unknown (as when stout Cortés and all his men, through Keats, in wild surmise gazed upon the Pacific).[2] In his discussion of Mercator's *Atlas* (1636), José Rabasa points out that although '[r]egions corresponding to *terra incognita* may lack precise contours' they are none the less presented in this book of maps within a framework already understood (in this case, on Rabasa's reading, a complex palimpsest of allegories): 'The *Atlas* thus constitutes a world where all possible "surprises" have been precodified' (1993, p. 194).[3] We do not feel the disruptions of space, the coming upon difference. On the road map you won't drive off the edge of your known world. In space as I want to imagine it, you just might.

The chance of space

For such a space entails the unexpected. The specifically spatial within time-space is produced by that – sometimes happenstance, sometimes not – arrangement-in-relation-to-each-other that is the result of there being a multiplicity of trajectories. In spatial configurations, otherwise unconnected narratives may be brought into contact, or previously connected ones may be wrenched apart. There is always an element of 'chaos'. This is the chance of space; the accidental neighbour is one figure for it. Space as the closed system of the essential section presupposes (guarantees) the singular universal. But in this other spatiality different temporalities and different voices must work out means of accommodation. The chance of space must be responded to.

So an argument for an element of chance in space chimes with the current *Zeitgeist*. That itself, however, may be more problematical than illuminating. It is popular today to revel in the glorious random mixity of it all. It is taken to be a form of rebellion against over-rationalisation and the dominance of closed structures. A reaction against some of the excesses and the one-sidednesses of 'the modern'. Too often, though, it is a weak and confused rebellion. For one thing, what may look to you like randomness and chaos may be someone else's order. The street market and the council estate are classic figures of contrast

for space • reorientations

here: the latter is bureaucratic, ordered, uniform (to be derided), the former humming with spontaneity. Or so we are constantly told. Jane Jacobs' *The death and life of great American cities* (1961) set the tone. Jonathan Glancey, musing upon the order/disorder conundrum, offers the thought that 'Disorder can, of course, produce variety, excitement and its own hit-and-miss beauty. ... those of us who cannot abide supermarkets ... love the messy vitality of street markets' (1996, p. 20). My heart is with him, but none the less ... urban street markets are in fact, as Jane Jacobs recognised, intricate constructions of multiple routines, rhythms, and well-worn paths – ordering systems. (To see them otherwise can resonate with elitist assumptions about the spontaneity of the life of the lower orders. And why is it anyway that while the uniformity of the council estate is always 'dreary uniformity' the bourgeois uniformity of Bath is universally celebrated? Could it be that the issue is not uniformity at all? There are all kinds of issues here; among them of class and politics.) What to me seems like the chaotic mess inflicted upon the city by deregulation and privatisation is proba-bly to those who have built their fortunes through it a game whose rules they know extremely well. It is 'the order of the market'. And again there is a poli-tics here. For while the order and uniformity which is rejected through so much easy critique is frequently associated with 'planning' or 'the state', the disci-plining order of the market or of other non-state social forces is more rarely subject to the same attention, hiding its power behind the new love affair with chaos (Wilson, 1991 comes close to this danger; for a corrective see Glancey, 1996). The use of the adjective 'state' as the iconic term of abuse in an era of cor-porate power can be dangerously misleading. As Lyotard (1989) argues, there is much in postmodern capitalism which coincides quite well with indetermi-nacy and the avant-garde sublime. Or again, Sadler (1998), writing of the situ-ationists, argues that the kind of architecture they endorsed 'existed by chance rather than design: backstreets, urban fabric layered over time, ghettos' (p. 159). It is the last of these which is particularly odd. What of the systematic and powerful ordering mechanisms of market and discrimination interlocked? So the language of order and chance has become loose and problematical. And yet it is important to emphasise that the element of surprise, the unexpected, the other, is crucial to what space gives us.

One way in which 'chance' has become integral to thinking about space is through architecture. Early situationists played with ideas in which buildings could be spaces which enabled the unexpected and the unplanned. Aldo van Eyck's Amsterdam Children's Home was designed as 'a place of chance encoun-ters and of the imagination' (Glancey and Brandolini, 1999, p. 16), and his sculp-ture pavilion at Arnhem was to have the effect of 'Bump! – Sorry. What's this? Oh hello!' (van Eyck, quoted in Jencks, 1973, p. 316; Sadler, 1998, p. 171), which captures beautifully the potential surprise of space. It is the accidental neighbour; the encounter with the unforeseen. What van Eyck was aiming at was a mixture of order and accident that he called 'labyrinthine clarity' (Sadler, 1998, p. 30).[4]

Such explorations continue, in particular perhaps in that architecture which is sometimes gathered together under the (often disputed) rubric of *deconstruction* (see, for instance, *Architectural Design*, 1988), and drawing too on a resonance of situationism. In the introduction in *Architectural Design* to the Academy Forum on Deconstruction at The Tate Gallery in 1988, Bernard Tschumi's architecture was described as addressing 'new concepts of space and time. ... Tschumi's aim is to challenge long celebrated icons and notions of the city to show that the city we inhabit is a fractured space of accidents' (p. 7). And later in the same issue Tschumi himself, discussing his *Folies* project for the Parc de la Villette, wrote, 'Above all, the project directed an attack against cause-and-effect relationships ... replacing these oppositions by new concepts of contiguity and superimposition' (Tschumi, 1988, p. 38). What was to be produced was 'something undecidable, something that is the opposite of a totality' (p. 38). Moreover, this undecidability resulted, not from some overall randomness, but through superimposing three separate structures (a point system, coordinate axes and a curve) each of which in themselves was coherently logical. Tschumi's argument was that superimposing these structures led to a questioning of 'their conceptual status as ordering machines: the superimposition of three coherent structures can never result in a super-coherent megastructure' (p. 38). It is the fact of spatial juxtaposition which produces the openness, the impossibility of closure into a synchronic totality. Or, to put it the other way around, this element of the chance/openness of space results from the co-existence of structures which are each in themselves by no means chaotic – it is the fact of multiplicity which produces the indeterminacy. Tschumi works towards an architecture which strives to be enabling of events (Tschumi, 2000a, 2000b). He writes of combinations 'of heterogeneous and incompatible terms', of juxtapositions of difference, of 'that event, that place of shock, or that place of the invention of ourselves' (2000a, pp. 174, 176). This surely captures something of the openness of spatiality. The imagery, however, is unfortunate. For Tschumi indeterminacy is produced through a layered horizontality. It is an indeterminacy which has its origins in the superimposition of three flat structures. The problem is that there is no temporality here. Space here is formed by putting together three closed horizontal surfaces.

I want to argue something different. Space is indeed 'undecidable' in Tschumi's sense, but that characteristic does not result from the superimposition of surfaces but from the spatial configuration of multiple (and indeed complex and structured) trajectories. Not the mutual interference of (horizontal) closed structures, but intertwined openended trajectories. In 'Six concepts' (2000a) Tschumi reflects upon the emergence of superimposition as a device within his approach to architecture. It was, he argues, a means of challenging the dualisms of form and function, structure and ornament, and the hierarchies implied within them. In a move which hints at a turn away from that horizontality of perspective that accompanies a focus on the discursive, he continues:

> Yet if I was to examine both my own work of this time and that of my colleagues, I would say that both grew out of a critique of architecture, of the nature of architecture. It dismantled concepts and became a remarkable conceptual tool, but it could not address the one thing that makes the work of architects ultimately different from the work of philosophers: materiality.
>
> Just as there is a logic of words or of drawings, there is a logic of materials and they are not the same. And however much they are subverted, something ultimately resists. *Ceci n'est pas une pipe.* A word is not a concrete block. The concept of dog does not bark. To quote Gilles Deleuze, 'The concepts of film are not given in film.' (p. 173)

This is a turn which bears a close relation to that shift of perspective which is entailed in moving from a concern with horizontalities to a focus on coeval trajectories.

But there are other sources too for the assumption of the significance of chance. One of them is 'Science'. The literature on chaos theory, complexity and uncertainty emanating from the natural sciences (originally meteorology – see Gleick, 1988), and most frequently with interpretative routes which have travelled through one or another understanding of quantum physics, is now used to license a celebration of undecidability in social matters too.

It is in this context that John Lechte (1995) has reflected on Breton and Tschumi and their relation to space. His concern is to explore the nature of 'postmodern space', in particular in relation to cities: 'architecture and the city are our concern' (p. 100) and 'we want to know what kind of space is constitutive of the postmodern city' (p. 102). And in this rethinking of the spatiality of postmodern cities the most crucial element Lechte highlights is undecidability: uncertainty, the element of chance. Surrealism is explored, and Derrida and deconstruction in architecture, and – inevitably – the *flâneur*. And towards the end of his article Lechte argues that, through indeterminacy, the element of chance renders space unrepresentable. It is an absorbing argument, and my much-read copy of the article bears the marks: this thought is underlined with definite approval.

And yet the manner of arriving at this conclusion raises further issues. Lechte begins from 'Science': 'developments in science are fundamental for helping us to understand what has happened in the modern (or the postmodern) city, and in particular what has happened in its architecture' (p. 100). His discussion of science follows familiar contours: that while nineteenth-century science was concerned above all with eliminating chance (this was the science of equilibrium and stasis), by the end of that century and into the twentieth the emergence of concepts of open systems and irreversible time led science itself to engage with and accept the fact of indeterminacy.[5] And this notion of indeterminacy in turn opens us up to 'a different understanding of the city. Postmodernity, I shall suggest, is, in part, this new understanding' (p. 102).

The first question concerns the general nature of Lechte's reliance on Science. He very interestingly argues the connection between certain developments in the natural sciences and the work of Lyotard, Derrida and Tschumi. Here he is writing about Lyotard's *The postmodern condition*: 'in this passage Lyotard is talking about science. He is not talking about politics or philosophy – least of all literary theory. I think that this is important because by limiting (but is it a limit?) himself to science, Lyotard is remaining within an area where there is still a good deal of consensus about the nature and importance of developments, even if these are poorly understood. Few people, for instance, would want to argue that quantum theory, or the theory of relativity, is ideologically charged' (1995, p. 99). Well. Ideological as opposed to …? (Think about current debates in biology.) Great shifts in the viewpoints of science are often imbricated with changes (and conflicts) in the society within which the scientific practice is embedded. There are huge debates about what quantum theory 'means', about how it should be interpreted (see, amongst many others, Bohm, 1998; Stengers, 1997) – indeed, Lechte's seems a rather unreflexive view in an article that is insisting on undecidability and the limits to knowledge.[6] It may be that the reliance on science should itself open up to a little undecidability.

However, there is also the question of what *kind* of chance is being referred to. It may be imagined in terms of the myriad of tiny causes which can contribute to any event – and this may be what Lechte is getting at when he writes of Bloom's walk in *Ulysses*: 'detail piled upon detail … until it seems impossible to take any more' (1995, p. 103). Then, the question is, is this a problem of our lack of knowledge (our inability to analyse) at such a level of minutiae? Or may it rather be interpreted as a real indeterminacy of process? At other points, Lechte picks up on a deconstructionist palimpsest understanding of chance (as in the case of Tschumi): a 'palimpsest image' where 'various levels … would show up "beneath" the surface of the standard version. This palimpsest quality renders determination fragile' (p. 106; here it is a Wittgenstinian notion of language which is being referred to). Or again, in a reinterpretation of Baudelaire's *flâneur* which moves away from a strictly modernist reading, Lechte writes:

> The *flâneur*'s trajectory leads nowhere and comes from nowhere. It is a trajectory without fixed spatial co-ordinates; there is, in short, no reference point from which to make predictions about the *flâneur*'s future. For the *flâneur* is an entity without past or future, without identity: an entity of contingency and indeterminacy. (p. 103)

How does this relate to postmodern science, to complexity and chaos theory – the sciences from which the article began? The connection certainly seems to be important to Lechte, who draws his argument through the aleatory wanderings of smoke and steam in Turner's paintings (see Serres, 1982). 'And in Turner's paintings wherein … lies randomness …? In smoke (steamships, locomotives, iron and steel foundries); … Thus would the very emblems of the modern industrial city give way to the indeterminacy which … makes for a different

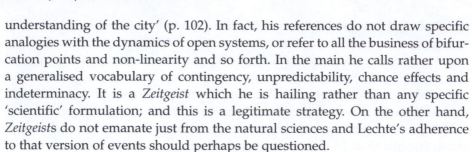

understanding of the city' (p. 102). In fact, his references do not draw specific analogies with the dynamics of open systems, or refer to all the business of bifurcation points and non-linearity and so forth. In the main he calls rather upon a generalised vocabulary of contingency, unpredictability, chance effects and indeterminacy. It is a *Zeitgeist* which he is hailing rather than any specific 'scientific' formulation; and this is a legitimate strategy. On the other hand, *Zeitgeist*s do not emanate just from the natural sciences and Lechte's adherence to that version of events should perhaps be questioned.

Moreover, this kind of general ontological uncertainty is not exactly what is at issue in the notion of the chance of space. This, though it may be part of the same broader phenomenon, is more specific. The chance of space lies within the constant formation of spatial configurations, those complex mixtures of pre-planned spatiality and happenstance positionings-in-relation-to-each-other that Tschumi was catching at. It is in the happenstance juxtaposition, in the unforeseen tearing apart, in the internal irruption, in the impossibility of closure, in the finding of yourself next door to alterity, in precisely that possibility of being surprised (the surprise which de Certeau argues is eliminated by spatialisation) that the chance of space is to be found. The surprise of space. And Lechte evokes this too: 'chance encounter upon chance encounter' (p. 103). But this is not unique to the postmodern city or peculiar to heterotopic spaces: all spaces are, at least a little, accidental, and all have an element of heterotopia. This is the instability and potential of the spatial, or at least of how we might in these space-times most productively imagine it.

It was something of this element of chance which situationist maps were trying to evoke. For them, among the characteristics of (urban) space was the resistance it necessarily offers to the homogenisation of the spectacle. The closure of space. But maybe the very impossibility of closing space, of reducing it to order (or even of 'conquering it'), gives hope that there is always a chance of avoiding recuperation – that there are always cracks in the carapace.

Yet chance alone is also insufficient; the *flâneur* is not enough to capture the city. Such images catch hold of only one side of things, and there is more to space than this. For 'chance', as Lechte himself points out, recalling Cournot's definition, may also be defined as 'the intersection of two or more chains of causality' (p. 110). There is chaos *and* order here. (Indeed, as Hacking (1990) points out, this 'long-standing idea of intersecting causal lines' is a 'face-saving, necessity-saving idea' which lies within a broader, deterministic understanding (p. 12).) The situationists disdained the surrealists' reliance on chance alone. Commenting on what he saw as the total failure of aimless surrealist ambulation, Guy Debord accused them sternly of 'An insufficient awareness of the limitations of chance, and of its inevitably reactionary use' (Debord, 1956/1981, cited in Sadler, 1998, p. 78), upon which Sadler comments that 'while situationists made it their business to disrupt the bourgeois worldview, they had no wish to problematize all instrumental knowledge and action' (p. 78). Or again, van Eyck's labyrinthine clarity, while like the situationists rejecting fixity and deterministic closure, was no collapse into total indeterminacy. Sadler aptly captures it as 'a more multifarious order' (p. 30). (And to take up again the iconic – if problematic – figure of the *flâneur*, Sadler records that, for all their rejection of the universalism of rationalist claims, for situationists and Team 10ers it was still 'Not that the drift of the pedestrian confounded all logic' (1998, p. 30).) Nor indeed are chance and indeterminacy the sole foci of any new science. Rather, there is the mutuality of chance and necesssity, and the Holy Grail for which many of the most ardent proponents of complexity are currently searching is 'deep order' (Lewin, 1993); order and disorder as folded into each other (Hayles, 1990; see also Watson, 1998).

Travelling imaginations

What is it to travel? How can we best think it in terms of time and space? Hernán Cortés trudging across the neck of (what was to become) Mexico. The 'voyagers of discovery' setting out across the oceans. My own, regular, journey to work: sitting in the train from London to Milton Keynes looking out of the window at the landscape we are crossing – out of the London basin, through the sharp gash carved in the chalk hills, emerging finally into the expanse of the clay of the East Midlands. Travelling across space? Is it? Thought of this way the very surface, of land or ocean, becomes equated with space itself.

Unlike time, it seems, you can see space spread out around you. Time is either past or to come or so minutely instantaneously *now* that it is impossible to grasp. Space, on the other hand, is *there*.

One immediate and evident effect of this is that space comes to seem so very much more *material* than time. Temporality seems easy to imagine in the abstract, as a dimension, as the dimension of change. Space, in contrast, has been equated with 'extension', and through that with the material. It is a distinction that resonates too (as was seen in Chapter 5) with that understanding of time as interior, as a product of (human) experience, in contrast to space as material *in opposition to* time's incorporeality: it is the landscape outside the window, the surface of the earth, a given.

There are many who have tried to puncture that smooth surface. The art events of Clive van den Berg (1997) aim to disrupt the complacent surface of white South Africa with reminders of the history on which it is based. Iain Sinclair's (1997) *dérives* through eastern London evoke, through the surface, pasts (and presents) not usually noticed. Anne McClintock's provocative notion of 'anachronistic space' – a permanently anterior time within the space of the modern – is catching at something similar (McClintock, 1995). On the way between London and Milton Keynes we go through Berkhamsted. Right by the station stand the remains of a Norman castle: the motte and bailey and

figure 11.2   *Berkhamsted Castle: past or present? (the ridge on the right is the railway embankment)* © Tim Parfitt

the moats around them still clearly defined, the grey stone walls now fallen and discontinuous, with the air of old grey teeth. We know then that the 'present-ness' of the horizontality of space is a product of a multitude of histories whose resonances are still there, if we would but see them, and which sometimes catch us with full force unawares.

However, it is not just buried histories at issue here, but histories still being made, now. Something more mobile than is implied by an archaeological dig down through the surfaces of the space of today. Something more temporal than the notion of space as a collage of historical periods (eleventh-century castle abutting nineteenth-century railway station).

So take the train, again, from London to Milton Keynes.[7] But this time you are not just travelling through space or across it (from one place – London – to another – Milton Keynes). Since space is the product of social relations you are also helping, although in this case in a fairly minor way, to *alter* space, to participate in its continuing production. You are part of the constant process of the making and breaking of links which is an element in the constitution of you yourself, of London (which will not have the pleasure of your company for the day), of Milton Keynes (which will; and whose existence as an independent node of commuting is reinforced as a result), and thus of space itself. You are not just travelling *through* space or across it, you are altering it a little. Space and place emerge through active material practices. Moreover, this movement of yours is not just spatial, it is also temporal. The London you left just a half an hour ago (as you speed through Cheddington) is not the London of now. It has already moved on. Lives have pushed ahead, investments and disinvestments have been made in the City, it has begun to rain quite heavily (they said it would); a crucial meeting has broken up acrimoniously; someone has caught a fish in the Grand Union canal. And you are on your way to meet up with a Milton Keynes which is also moving on. Arriving in a new place means joining up with, somehow linking into, the collection of interwoven stories of which that place is made. Arriving at the office, collecting the post, picking up the thread of discussions, remembering to ask how that meeting went last night, noticing gratefully that your room's been cleaned. Picking up the threads and weaving them into a more or less coherent feeling of being 'here', 'now'. Linking up again with trajectories you encountered the last time you were in the office. Movement, and the making of relations, take/make time.

At either end of your journey, then, a town or city (a place) which itself consists of a bundle of trajectories. And likewise with the places in between. You are, on that train, travelling not across space-as-a-surface (this would be the landscape – and anyway what to humans may be a surface is not so to the rain and may not be so either to a million micro-bugs which weave their way through it – this 'surface' is a specific relational production), you are travelling *across trajectories*. That tree which blows now in the wind out there beyond the train window was once an acorn on another tree, will one day hence be gone. That field of yellow oil-seed flower, product of fertiliser and European subsidy, is a moment – significant but passing – in a chain of industrialised agricultural production.

There is a famous passage, I think from Raymond Williams … He too is on a train and he catches a picture, a woman in her pinny bending over to clear the back drain with a stick. For the passenger on the train she will forever be doing this. She is held in that instant, almost immobilised. Perhaps she's doing it ('I really *must* clear out that drain before I go away') just as she locks up the house to leave to visit her sister, half the world away, and whom she hasn't seen for years. From the train she is going nowhere; she is trapped in the timeless instant.

Thinking space as the sphere of a multiplicity of trajectories, imagining a train journey (for example) as a speeding across on-going stories, means bringing the woman in the pinny to life, acknowledging her as another on-going life. Likewise with Berkhamsted Castle. The train does not, as some argue, speed across different time-zones, from Norman times to twentieth century. That would be to work with a form of theatre of memory which understands space as a kind of composite of instants of different times, an angle of the imagination which is a-historical, working in opposition to a sense of temporal development. Space as a collage of the static. Yet both the castle and the station continue their histories as I pass through (I may contribute to those histories). From Norman stronghold, the castle became a palace, was passed between kings and other royalty, served as a prison and was subsequently cannibalised for the building of a mansion. Today its story continues as a significant tourist attraction. (However much the heritage industries might wish on occasions to preserve things in aspic they cannot actually ever hold them still. The depthless commodified present which Jameson so effectively points to precisely denies all this. But it does so not only, as is usually argued, by commodifying 'the past', but also by refusing to recognise the histories which are ongoing through the present.) 'The only adequate image is one that includes a sense of motion in itself' (Rodowick, 1997, p. 88). The train transects the castle's on-going history.

As Jameson argued (Chapter 7), recognising all this is impossible. Every train journey (and that would be the least of it) would become a nightmare of guilty admission of all the stories the fullness of whose coeval existence you did not manage to recognise ... as the train sped on. What is at issue is not this but the change in perspective ... the imaginative opening up of space. It is to refuse that flipping of the imaginative eye from modernist singular temporality to postmodern depthlessness; to retain at least some sense of contemporaneous multiple becomings.

When Hernán Cortés heaved to the top of the pass between the snow-covered volcanoes and looked down upon the incredible island city of pyramids and causeways, the immense central valley between the mountain ranges stretching away into the heat, he wasn't just 'crossing space'. What was about to happen, as he and his army, and the discontented locals they had recruited along the way, marched down upon Tenochtitlán, was the meeting-up of two stories, each already with its own spaces and geographies, two imperial histories: the Aztec and the Spanish. We read so often of the conquest of space, but what was/is at issue is also the meeting up with others who are also journeying, also making histories. And also making geographies and imagining space: for the coeval look back, ignore you, stand in a different relation to your 'here and now'. Conquest, exploration, voyages of discovery are about the meeting-up of histories, not merely a pushing-out 'across space'. The shift in naming, from *la conquista* to *el encuentro,* speaks also of a more active imagination of the engagement between space and time. As Eric Wolf (1982) has so well reminded us, to think otherwise is to imagine 'a people without history'. It is to immobilise – suspended awaiting our arrival – the place at the other end of the journey; and it is to conceive of the journey itself as a movement simply across some imagined static surface.

Wolf's arguments, and the writings of others in a similar vein, are now well recognised and widely cited. Yet their implications are rarely taken on board; and this failure has political effects. José Rabasa's appreciative but critical engagement with the work of Michel de Certeau provides a lovely illustration both of how a contrary way of thinking (that 'others' 'out there' have no history) is still deeply embedded in the way we imagine the world and of why this matters. Rabasa (1993) analyses in particular de Certeau's treatment of Jean de Léry's *Histoire* of his journey in Brazil (de Certeau, 1988; de Léry, 1578), and draws out the opposition which de Certeau establishes in de Léry between two 'planes'. He quotes:

> On the first is written the chronicle of facts and deeds … These events are narrated in a *tense*: a *history* is composed with a chronology – very detailed – of actions undertaken or lived by a *subject*. On the second plane *objects* are set out in a space ruled not by localization or geographic routes – these indications are very rare and always vague – but by a taxonomy of living beings, a systematic inventory of philosophical questions, etc.; in sum, the catalogue raisonné of a knowledge. (de Certeau, 1988, pp. 225–6; cited in Rabasa, 1993, pp. 46–7; emphasis in the original)

de Certeau is here establishing a set of oppositions: between an active historical Europe and a passivity-to-be-named; between an agency/subject and an object of the gaze/knowledge; and (though Rabasa does not comment on this) between time and space. Rabasa's first point mirrors the arguments already made (Chapter 3) which are critical of de Certeau's 'insistence on binarism' (Rabasa, 1993, p. 46), and relates this to de Certeau's roots within structuralism and 'the danger of repeating the categories of the method under criticism' (p. 43) – the difficulty, even in critique, of fully escaping its terms.

But Rabasa then goes further. The 'passivity' was in fact not simply passive, he argues; Brazil was *not* simply an object of knowledge. As in Latin America more widely there was a substantial input to the colonial interpretation of this 'new world' from active indigenous knowledges. This was not 'Western desire' striding into the 'blank page' of the to-be-conquered/colonialised: rather, and however unequal were the terms, it was an encounter. (In the language of the argument of this book, there was more than one history here.) Moreover, argues Rabasa, it is not only in terms of an interpretation of the past that such binary readings have effects: more generally they construct a tautological closure which ignores a potential openendedness; it is a 'will to closure' which must be prised open precisely to enable a way out from present-day Eurocentrism.

Now, what Rabasa does not do (it was not his concern) is to pull out what is going on here in terms of time and space. This, too, is an opposition embedded in the quotation from de Certeau (although it should be recognised that the possibility is also suggested that space can be traced through 'routes' – that it can be more active, mobile?). In this formulation history/time is the active term, voyaging across passive geography/space. It is thus that the 'others' are rendered static, without history.

It is thus, too, that they can be rendered as 'a blank page'. This is a significant phrase: one deployed by de Certeau and analysed by Rabasa, and it links us back to other themes. Rabasa's argument is that the construction and interpretation of these active/passive discourses of colonialism (and, in my terms, these discourses of time and space) are bound up with wider historical shifts.

In the first place, they are bound up with a more generally emerging distinction between a 'subject' and an 'object' of knowledge (and, in Rabasa's view, with 'the emergence of Western subjectivity as universal') (p. 47). Secondly, they are bound up with the emergence of 'the scriptural economy of the Renaissance' and the strict distinguishing of writing from orality, with the latter designated as the primitive form: 'it is only in the Renaissance that writing defined itself as labor, in opposition to non-productive orality. This scriptural economy reduced Amerindians to "savages" without culture, hence to apprentices of Western culture' (pp. 51–2). Orality is banished to the spatiality of the object; one writes *on* it. (Just as one, supposedly, travels *across* space.)

Now, both the term 'the scriptural economy of the Renaissance' and Rabasa's link between orality and spatiality are drawn from de Certeau (de Certeau, 1984, ch. 10; and 1988, ch. 5, respectively).[8] De Certeau writes, 'The "difference" implied by orality … delimits an *expanse of space,* an object of scientific activity. In order to be spoken, oral language waits for a writing to circumscribe it and to recognise what it is expressing' (de Certeau, 1988, p. 210; emphasis in the original). Two uses thus come together: the blank page of what will become, in this case, the Americas 'on which Western desire will be written '(1988, p. xxv) and the blank page as 'the proper place of "writing"' (Rabasa, 1993, p. 42). For de Certeau, 'writing' is 'the concrete activity that consists in constructing, on its own blank space (*un espace propre*) – the page – a text that has power over the exteriority from which it has first been isolated' (de Certeau, 1984, p. 134). The notion of a blank page relates both to the conceptualisation of 'the "Other" as absence of culture' (Rabasa, 1993, p. 42) – or in my terms and more generally as absence of history/trajectory – *and* to the connection between writing-as-representation and space. And, as will be remembered from Chapter 3, for de Certeau 'The "proper" is a victory of space over time' (1984, p. xix). Moreover, as Rabasa goes on to argue, in relation to the development of the printing press in contrast to 'the scribes of the Middle Ages', 'books and maps … not only made information more accessible but also laid out the world on *surfaces* ready to be "explored"' (1993, p. 52; my emphasis).[9]

Two things are working together here then, and they powerfully reinforce each other. On the one hand the representation of space as a surface, and on the other hand the imagination of representation (here, again, in the specific form of writing, as scientific representation) in terms of spatialisation. Together what they lead to is the stabilisation of others, their deprivation of a history. It is a political cosmology which enables us in our mind's eye to rob others of their histories; we hold them still for our own purposes, while we do the moving. Crucial to this operation is the taming of space.

And here this argument can link up with others. For we perform such magic with our usual notions of space. Not only do we imagine it as a surface, we do in fact often conceive of our journeys 'across' it as temporal too. But not the way I mean it, where our trajectory will meet up with another's. As has been argued, 'the West', in its voyages and in its anthropology, and in its current imaginings of the geography of globalisation, has so often imagined itself going out and finding, not contemporary stories, but the past. (Do travellers to California imagine themselves as accelerating through history?) Or, again, there is the way the story of cities is so often told, as a tale of singular change from Athens to Los Angeles. (Where in this line of development do we put Samarkand or São Paulo? Does it mean Calcutta will one day be like LA? And what of Bangalore?) Space as a surface, then, but one which slopes in time.

We do it in our daily lives. Migrants imagine 'home', the place they used to be, *as* it used to be. The 'Angry Young Men' of the British 1950s and 1960s have become iconic in this; coming south to make their names, both ridiculing and, so often in the figure of 'Mother', sometimes revering, the northern places they had left. But what they so often also tried to do was hold those places in aspic; they stopped these places' histories at the point at which the migrants left. The spatial surface, from London to the north, sloped backwards in time.

I too am a northerner who presently lives 'down south' and I have often thought about this in the context of 'going home'. When the train passes Cloud Hill beyond Congleton we're nearly there. I put away my books (this is a ritual), the hills get higher, the people get smaller, and I know that when I get off the train I will meet again the constant cheery back-chat which is south Lancashire. I'm 'home', and I love it, and part of what I love is my richer set of connections here, precisely its familiarity.

And what is wrong with that? This kind of longing, for instance of the migrant, for a 'home' they used to know? Wendy Wheeler (1994) has addressed this question in her thoughtful work about the losses we have suffered as a price of our incorporation in the project of modernity (see also Wheeler, 1999). As do many others, she points to the prominence within the postmodern of feelings and expressions of *nostalgia,* including nostalgias for place and home (one section is entitled: 'postmodernity as longing to come home'). While agreeing that the fixing of the identity of places is a matter always of power and contestation rather than of actually existing authenticity, and agreeing too that 'the past was no more static than the present' (she is citing, and responding at this point to, Massey, 1992b, p. 13), she continues, 'it is nevertheless still the case, as Angelika Bammer argues (Bammer, 1992, p. xi), that these nostalgic gestures of postmodernism are "the recuperative gestures of our affective needs". One of the questions which postmodernism poses to politics is that of a response to "affective needs"' (Wheeler, 1994, p. 99). Her argument is that Enlightenment modernity has been bought at the cost of the radical exclusion of everything that might threaten rational consciousness. Moreover,

This radical exclusion of Reason's 'other' forms the basis both of the major distinctions upon which modernity is founded (reason/unreason; maturity/

childishness; masculinity/feminity; science/art; high culture/mass culture; critique/affect; politics/aesthetics etc.) and of modern subjectivity itself. (p. 96)

This is an important argument, and one which in a number of ways links up with the theses in this book.[10] Postmodern nostalgia, on this reading, is at least partly explicable as a kind of return of the repressed of modernity. Moreover, it can take a number of forms, and one potential political project is precisely to articulate a politically progressive form. The title of Wendy Wheeler's article is 'Nostalgia isn't nasty'.

Now, nostalgia constitutively plays with notions of space and time. And what I would like to argue, I think in sympathy with Wheeler's thesis at its broadest level, is that when nostalgia articulates space and time in such a way that it robs others of their histories (their stories), then indeed we need to rework nostalgia. Maybe in *those* cases it is indeed 'nasty'.

My point is that the imagination of going home (and I am by no means sure that, as Wheeler implies, this is only a postmodern phenomenon) so frequently means going 'back' in both space and time. Back to the old familiar things, to the way things used to be. (Indeed as I look out the window after Congleton the things I pick out are so often the things I remember from before. Signs of Mancunian specificity, which so often too get entangled (given modernity's *and* postmodernity's tendencies to sameness) with signs inherited from the past – one thinks wryly of Borges' (1970) 'The Argentine writer and tradition'.)

One moment haunts me in this regard. My sister and I had gone 'back home' and were sitting with our parents in the front room having tea. The treat on such occasions was the chocolate cake. It was a speciality: heavy and with some kind of mixture of butter, syrup and cocoa powder in the middle. A wartime recipe I think, invented out of necessity, and a triumph. I loved it. On this occasion, though, Mum went out to the kitchen and came back holding a chocolate cake that was altogether different. All light-textured and fluffy, and a paler brown. Not the good old stodgy sweetness we loved so well. She was so pleased; a new recipe she'd found. But with one voice my sister and I sent up a wail of complaint – 'Oh *Mum* … but we like the *old* chocolate cake'.

I've often re-lived and regretted that moment, though I think she understood. For me, without thinking then of its implications, part of the point of going home was to do things as we'd always done them. Going home, in the way I was carrying it at that moment, did not mean joining up with ongoing Mancunian lives. Certainly it was time travel as well as space travel, but I lived it in that moment as a journey to the past. But places change; they go on without you. Mother invents new recipes. A nostalgia which denies that, is certainly in need of re-working.

For the truth is that you can never simply 'go back', to home or to anywhere else. When you get 'there' the place will have moved on just as you yourself will have changed. And this of course is the point. For to open up 'space' to this kind of imagination means thinking time and space as mutually imbricated and thinking both of them as the product of interrelations. You can't go back in space-time. To think that you can is to deprive others of their ongoing independent stories. It may be 'going back home', or imagining regions and countries as backward, as needing to catch up, or just taking that holiday in some 'unspoilt, timeless' spot. The point is the same. You can't go back. (De Certeau's trajectories are not, in fact, reversible. That you can trace backwards on a page/map does not mean you can in space-time. The indigenous Mexicans might re-trace their footsteps, but their place of origin will no longer be the same.) You can't hold places still. What you *can* do is meet up with others, catch up with where another's history has got to 'now', but where that 'now' (more rigorously, that 'here and now', that *hic et nunc*) is itself constituted by nothing more than – precisely – that meeting-up (again).

(A reliance on science? 3)

I have argued that there is a particular kind of mix of order and chance that is integral to the continual process of spatial (re)configuration in an open space-time; the loose ends, the elements of chaos, the meetings without merging.

There are strategic reasons for proceeding in this particular way. To attempt to ground these arguments by a general gesture to, for instance, chaos theory or complexity theory, quite apart from hedging on arguments concerning the ontological assumptions implicit in such claims, would both downgrade the point I am wanting to make and lose sight of the specificity of the mechanisms I wish to point to. Moreover, subsuming the specifically spatial characteristic of openness and indeterminacy within some general reference to the (generally currently accepted) complexity and indeterminacy of just about everything, would lose the ability also to point to the social scientific and political implications of taking seriously the specificity of the chance of space.

None the less, it would be disingenuous to deny any connection between the debates about spatiality and the wider circulation of ideas about complexity and indeterminacy. Indeed, it is arguable that what has been going on is not simply the adoption and generalisation by social scientists and philosophers of ideas which have their ultimate origin in a natural science of which those social theorists are in awe. Thus Nigel Thrift (1999) argues that ideas of complexity have come to frame 'a commonplace structure of intelligibility' (p. 35; emphasis in the original) and that complexity theory 'might be seen as one of the harbingers of ... the emergence of a structure of feeling in Euro-American societies which frames the world as complex, irreducible, anti-closural and, in doing so, is producing a much greater sense of openness and possibility about the future' (p. 34; my emphasis). For Thrift, 'the metaphors of complexity theory are both a call and a response' (p. 53) to this emerging structure of feeling.[11] This is a helpful reconfiguration of what is going on. The specifics of complexity theory are themselves embedded in a wider Zeitgeist.

This resituation raises further considerations. First, there is the argument (Part Two) that the routes travelled by ideas are complex and multidirectional. The Zeitgeist does not have singular roots in a particular domain of thinking, such as the complexity theory of natural science. The passages of concepts, and the translations and transformations which occur on the way, are likely to be multifarious (Thrift, 1999). Zohar, indeed, reverses what is perhaps the more common assumption and argues that 'Like Newtonian science before it, twentieth-century science has grown out of a deep shift in general culture, a move away from absolute truth and absolute perspective toward contextualism; a move away from certainty, toward an appreciation for pluralism and diversity, toward an acceptance of ambiguity and paradox, of complexity rather than simplicity' (1997, p. 9; my emphasis). And indeed, rather differently, Thrift hypothesises that complexity theory might well be being more successfully propagated outside natural sciences than within. This labyrinthine nature of the travelling of ideas is of course a more general phenomenon. Prigogine and Stengers (1984, especially Chapter 1) place their argument firmly in the context of a long historical exchange between natural sciences on the one hand and philosophy/social sciences on the other. Stengers, whose wider position is to argue both for greater communication between science and philosophy and for greater scepticism about the authority of science, produces a highly nuanced consideration of the potential and the dangers inherent in the voyages of this particular idea (Stengers, 1997, especially Chapter 1, which is entitled 'Complexity: a fad?'). Deleuze (1995), when questioned about his own use of concepts from contemporary physics, referred precisely to Prigogine and proposed that the concept of bifurcation is 'a good example of a concept that's irreducibly philosophical, scientific, and artistic too' (pp. 29–30). Philosophers may create concepts that are useful in science and, most importantly, 'no special status should be assigned to any particular field, whether philosophy, science, art, or literature' (p. 30).

It may, then, be more appropriate to interpret references to complexity theory, even when as in Lechte's case they appeal quite explicitly to a natural science as a legitimising ground for their argument, rather as particular elements in a wider and multiply-interconnected structure of intelligibility which is emerging as common to the age, at least in certain Western countries. None the less, I would argue, we are still duty-bound to address ourselves to a number of more particular questions. Thus, I would maintain, we still have to specify, each in our own field of study, just what we mean by hailing this general reference into our particular area, and just what work it does, upon what issues it gives us more effective purchase. This question emerges as a fascinating thread of debate in Lewin (1993).

Moreover, and this is the most important point, there is anyway no necessity to go along with the Zeitgeist. Of every Zeitgeist, every structure of feeling, which we hail and employ, it is surely necessary to ask: is this in tune, not just with 'the times' (so what?), but with how we wish (socially, politically) to address these times? It may be that we wish precisely to subvert the dominant cultural tendencies of the moment.

However, there is perhaps a more precise connection, which goes beyond a generalised resonance, between concepts of complexity on the one hand and a re-evaluation of the significance of space on the other. It is frequently argued, for instance, that in the most general of terms the theory of complexity evokes 'the spatial', that what it is all about is the kind of spatial configurations that are provoked by the channelling of energies. Certainly, the whole notion of distributed systems, the practices of parallel processing, and even the idea of emergence itself, necessarily carry within them implications of multiplicity as opposed to a singular linearity. They precisely depend on complex interrelationality. And multiplicity and interrelatedness in turn entail, in the argument being presented here, spatiality. (This does not mean, even so, that we should turn to complexity theory for justification for such views. Feminists working towards relational thinking got there by different routes, those imagining the emergence of identity out of multiplicity did so likewise … and I would argue the same for our thinking around spatiality.) In relation to the particular connection between complexity and spatiality, Thrift writes: 'Whereas previous bodies of scientific theory were chiefly concerned with temporal progression, complexity theory is equally concerned with space. Its whole structure depends upon emergent properties arising out of excitable spatial orders over time' (1999, p. 32). But again we must be careful, for there are a number of different steps here. As Part Two was at pains to show, and as those theorists most concerned to propagate the implications of complexity theory insistently argue (Stengers, Prigogine), 'previous bodies of scientific theory' were in fact on their own readings precisely abstracting from *the historical messiness the reassuringly stable (for them 'spatial') eternal truths. I would argue, then, rather differently: that if there is this general connection between complexity theory and spatiality it is also because the former has the potential to force the latter to mean something different. No longer can 'space' be the ultimate pinning-down and stabilisation, through scientific representation, of the fundamental laws of the world. Rather, spatial configuration is now interpreted as a significant factor in the emergence of the new. It is not, then, that space, in unchanged meaning, suddenly finds itself put upon the stage, but rather that what we mean by space has also been (or is potentially) revolutionised.*

There are, moreover, particular aspects of complexity theory which resonate with this potentially revolutionised imagination of space. There is an emphasis on juxtaposition, on encounter and entanglement and on their not-always-predictable effects: on the configurational. And above all there is, on some readings of complexity theory at least, an insistence on the understanding of temporality as open. So if such connections exist, if the indeterminations of complexity resonate with the indeterminations which arise when a (reimagined) spatiality is integrated more fully into our analyses, then this could be another element of the current Zeitgeist which accounts for what has been called the 'spatial turn' in social theorising.

And yet the dimensions of that connection remain largely unrecognised or, at least, are frequently implicit. There is a further element in the implications held out by the burgeoning networks of the metaphor of complexity. For few of those who write about complexity, and who engage in this natural science/social science cross-talk, take the argument as far as the implications it holds out for how we think about space. Isabelle Stengers, for instance, one of the key reference points in all of this, is meticulous and thought-provoking about time; but she doesn't mention space. In her collection Power and invention: situating science *(1997) there are nineteen entries in the index to 'time', with a trail of subheadings and a cross-reference; there is not a single entry for 'space'. The idea of complexity, she argues, is intimately tied up with 'that singular category of objects that must be called historical' (p. 13). A number of paths are then pursued in an elaboration of mechanisms which constitute this historical nature (that is, the temporal irreversibility) of such objects. One of these paths concerns memory; in other words, one of the elements producing irreversibility is memory and the associated possibility of learning. And Stengers evokes 'the memory of all the pasts' (p. 17) which make possible such learning processes, and which in turn mean that the future will not just be a reiteration of the past. Likewise she evokes, as another path, the notion of context, and this is glossed as 'being produced by history and capable of history' (p. 17). 'Pasts' and 'histories'. Both temporal. But memories and contexts are also spatial. So I would add, to pasts and histories, 'elsewheres' and 'geographies' too.*

Now, of course, it is possible to reply that the past is assumed to be placed and that 'history' of course is meant to include geography. It's implicit. Too obvious to mention. But this is just my point: by leaving space implicit one fails to draw out both the import of this tremendous argument about irreversibility for how we think about space itself and the particular aspect in our imagination of space-time that this reconceptualised spatiality can highlight. For in the context of (at least until recently) hegemonic understandings of memory, the most likely connotations are to the internalised individual, and the notion of history may well be the singular history. Highlighting the spatiality of our pasts and the geography of our histories – the dispersion of our very selves – entails a more outward-looking understanding in which all these things are necessarily constituted in and through contacts, relations, interconnections, with others.

Such an outward looking, relational, understanding is of course basic to the way in which Stengers thinks. The whole notion of context in her sense implies the multiplicity which is essential for historicity. Thus,

> *a bird, a chimpanzee, or a human being learns. The behaviour of the individual does not repeat the species since each one constitutes a singular construction that integrates genetic constraints and the circumstances of a life. Furthermore, selective pressure does not bear on the individual but on the individual in its group, in the strong sense … The group has become the condition of possibility for the individual, whose development involves protection, learning, and relations. (p. 16; emphasis in the original; Marx would approve)*

She goes on, 'The individual now appears as a sheaf of linked temporalities' (p. 16; my emphasis). This is wonderful stuff. The logic, though, could be pushed just one step further. For what Stengers is arguing for is the recognition by scientific practice of this essential element of historicity (such as comes with processes of learning). However, not only in order to have such an open historicity do you need an open and relational space but also such a notion of space is quite the opposite of that language of spatiality (where space = static representation = the obliteration of temporality) which surrounded the physics of reversibility. It is not only the understanding of time which this argument challenges but, potentially, also the understanding of space.

12
the elusiveness of place

Migrant rocks

One way of seeing 'places' is as on the surface of maps: Samarkand is *there*, the United States of America (finger outlining a boundary) is *here*. But to escape from an imagination of space as a surface is to abandon also that view of place. If space is rather a simultaneity of stories-so-far, then places are collections of those stories, articulations within the wider power-geometries of space. Their character will be a product of these intersections within that wider setting, and of what is made of them. And, too, of the non-meetings-up, the disconnections and the relations not established, the exclusions. All this contributes to the specificity of place.

To travel between places is to move between collections of trajectories and to reinsert yourself in the ones to which you relate. Arrived at work, in Milton Keynes, I rejoin debates, teams meeting to discuss teaching, a whole cartography of correspondence, ongoing conversations, pick up where I left off the last time I was 'here'. Back in London at night I emerge into the energising bedlam of Euston Station and go through the same process again. Another place, another set of stories. I catch the headlines on the *Evening Standard* (what's been going on?). Leaving the station, I search the sky and the pavements, wondering what the weather's been like (will my garden be crying out for water?). Finally, arrived back in my flat, I check the post, the telephone messages, find out 'what's been happening here' while I've been away. Bit by bit I reimmerse myself into (just a few of) the stories of London. I weave together the stories which make this 'here and now' for me. (Others will weave together different stories.) Sometimes there are attempts at drawing boundaries, but even these do not usually refer to everything: they are selective filtering systems; their meaning and effect is constantly renegotiated. And they are persistently transgressed.[12] Places not as points or areas on maps, but as integrations of space and time; as *spatio-temporal events*.

This is an understanding of place – as open ('a global sense of place'), as woven together out of ongoing stories, as a moment within power-geometries, as a particular constellation within the wider topographies of space, and as in process, as unfinished business – which I have often written about before (Massey, 1991a, 1997a, 2001a). To all of which a friend has over the years persistently replied, 'That's all right when you talk about human activity and human relations. I can understand and relate to it then: the interconnectivity, the essential transience … but I live in Snowdonia and my sense of place is bound up with the mountains.'[13]

Some of our strongest evocations of place (in the Western world but not only there) indeed draw on hills, on 'the wilderness' (dubious category anyway), on the sea. We escape from the city maybe to replenish our souls in contemplating the timelessness of mountains, by grounding ourselves again in 'nature'. We use such places to situate ourselves, to convince ourselves that there is indeed a grounding. It recalls too, however, that untenable disjunction between the celebration of cultural flow and mixity and the nervousness at a natural world that will not stay still, which was remarked upon in Chapter 9. How then to think this notion of place as a temporary constellation, as a time-space event, in relation to this 'other' arena, 'the natural world'?[14]

My imagination was reworked some winters ago, while in the northern Lake District, in north west England. It would be easy to write of the Lake District, or of Keswick, the town where I was staying with my sister, as a bundling of different social stories with different spatial reaches and differing temporalities. Longstanding farmers, the grey-stone country houses of the aristocratic incomers of the eighteenth and nineteenth centuries, poets and Romanticism, ancient mining, middle-class cottage owners, Roman remains, an international tourist trade, a focus of a discourse of the sublime … But just outside the town looms Skiddaw, a massive block of a mountain, over 3000 feet high, grey and stony; not pretty, but impressive; immovable, timeless. It was impossible not to consider its relationship to this place. Through all that history, it seemed, it had presided.

It is evident, of course, that much of the landscape here has been etched and moulded into its present-day basic shape by the glaciers of ice ages, the last of which retreated some 10,000 years ago. The traces are everywhere: in the U-shaped valleys inherited and reused in the last advance of the ice, in the hummocky landscape of moraines (material dumped by ice as it passed), in so-called *roches moutonnées* (rocks which have been scraped smooth and striated as the ice ground over them then plucked into jagged shapes on the downstream – downglacier – side), and in drumlins, of which there are many in these parts, egg-shaped hills deposited under the ice as the glacier passed on and over, from what is now the valley of Derwentwater north to

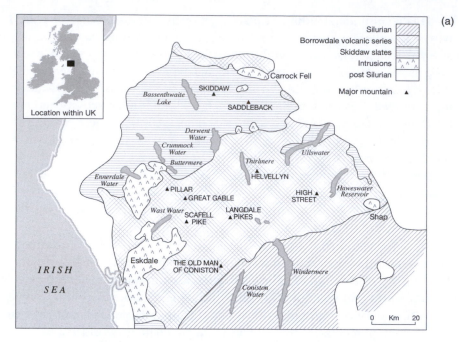

(a)

figure 12.1a   *Simplified geology of the Lake District (after Goudie and Sparks)*

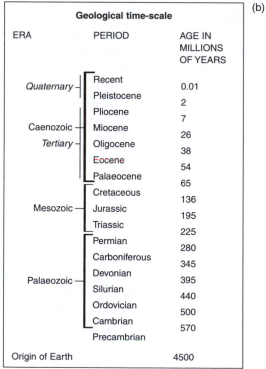

(b)

Geological time-scale

ERA	PERIOD	AGE IN MILLIONS OF YEARS
Quaternary	Recent	0.01
	Pleistocene	2
Caenozoic	Pliocene	7
	Miocene	26
Tertiary	Oligocene	38
	Eocene	54
	Palaeocene	65
Mesozoic	Cretaceous	136
	Jurassic	195
	Triassic	225
	Permian	280
	Carboniferous	345
Palaeozoic	Devonian	395
	Silurian	440
	Ordovician	500
	Cambrian	570
	Precambrian	
Origin of Earth		4500

figure 12.1b   *Geological time series*

Bassenthwaite. The hotel where we were staying stands on a graciously sweeping road which takes its shape not just from some designer's preference for curvacious avenues, but from following the foot of a drumlin. Ancient ice ages plainly readable in the human landscape. One thing it might evoke is the antiquity of things. But another is almost the converse: that today's 'Skiddaw' is quite new.

I knew, too, that the rocks of which Skiddaw is made were laid down in a sea which existed some 500 million years ago. (They are composed from the erosion of still older lands.) And 'not long' afterwards (in the same – Ordovician – geological period) there was volcanic activity. There are reminders of that tumultuous era too in the present-day landscape. Today's mountains bear no relation to the ancient volcanoes, but these more resistant volcanic rocks to the south give rise to a markedly different scenery of cliffs and waterfalls. And for those who know how to spot them, there are outcrops of lavas and tuffs. Some volcanic rocks form the cores of drumlinshaped hills: the remnants of volcanic activity from over 400 million years ago, plastered millions of years later by debris deposited by the retreating glacier (Boardman, 1996). A long and turbulent history, then. So much for 'timelessness'.

Such observations are not so startling. (Two hundred years ago, before geologists such as Charles Lyell, they would have been shocking if not incomprehensible. The opening up by geology and palaeontology of that deep history challenged prevailing notions of time, shook established Judaeo-Christian religious thinking … and made possible a different reading of landscape and place.) Reading history in the rocks is not so revelatory today. Even Baudrillard refers to 'the remorseless eternity' of geology (1988, p. 3) as he belts across the 'American' desert (though he doesn't do much with it, doesn't explore how it could challenge (rather than confirm) the notion of depthlessness, just as his use of the term 'America' ignores the history of that name and his complicity in its appropriation by the USA alone). What this geological history tells us is that this 'natural' place to which we appeal for timelessness has of course been (and still is) constantly changing.

But it's not merely a question of time: that history had a geography too. Sitting in our room at night, hemmed in by the (apparent) steadfastness of nature in the dark outside, and poring over local geology, the angle of vision shifted. For *when* the rocks of Skiddaw were laid down, about 500 million years ago, they were not 'here' at all. That sea was in the southern hemisphere, about a third of the way south from the equator towards the south pole. (Rude shock this, for Skiddaw is a mountain which, in English imaginations, is inextricably of 'the North'. I grew up singing 'Hills of the North rejoice'.)

Geological imaginations have their histories too, of course; what follows is what I understand of currently hegemonic ones.[15] On the planet on which

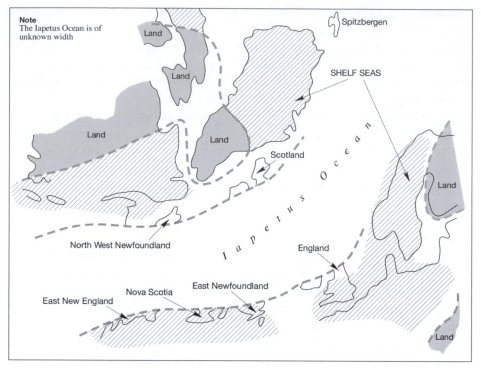

figure 12.2   *The Iapetus Sea: where the Skiddaw slates were laid down (after Windley and Cowey)*

this sea existed, where the slates were deposited, floated various bits and assemblages of the continents which we have today. The sea is now (that is, by current geologists, tectonicists et al.) called Iapetus, and it lay between two of these ancient continents (the volcanic activity was sparked off as they moved). The whole thing has subsequently floated about the planet as the continents rearranged themselves. The bit that we know today as the slates of Skiddaw crossed the equator about 300 million years ago. (And this in turn was way before 'the Americas', although of course they were not called that then – there were still 450 million or so years to go before Hernán Cortés would cross the Atlantic and Amerigo Vespucci would be born – were beginning to break away from the great old rock plateaux of what we now call southern Africa. Anyway, it was only relatively recently that there began to *be* an Atlantic for Hernán Cortés to cross.) And it was a mere 10 million years ago that the rocks of the present-day mountain rose above the surface of the ocean. The 'history' represented in the geological series in figure 12.1b erases a mobile geography. And it wasn't as though I hadn't 'known' all this; what startled was the shift in imagination – the real *appreciation* of it.

Nor was this yet in the shape of what we might propose as 'a mountain' (Latour, 2004), still less one called Skiddaw. That took, as the rocks were moving

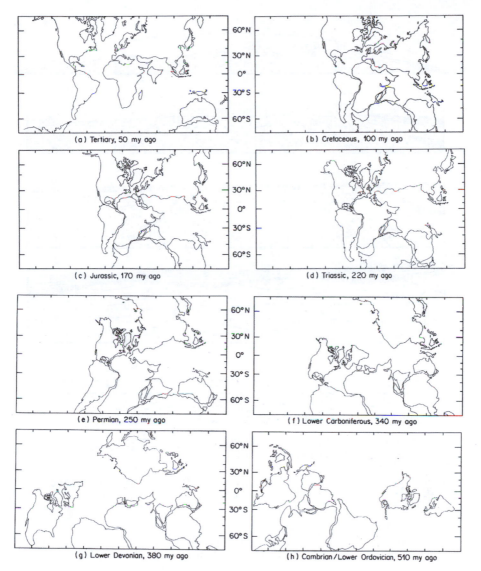

(a) Tertiary, 50 my ago

(b) Cretaceous, 100 my ago

(c) Jurassic, 170 my ago

(d) Triassic, 220 my ago

(e) Permian, 250 my ago

(f) Lower Carboniferous, 340 my ago

(g) Lower Devonian, 380 my ago

(h) Cambrian/Lower Ordovician, 510 my ago

figure 12.3   *Continental drift from the Cambrain to the Tertiary (after Smith Briden and Drewry, 1973)*

Source: © The Palaeontological Association

northwards, great periods of folding and contortion, injections of igneous rocks from below, periods of differential erosion, overlay by other strata and their folding and denudation, shifts in altitude.

When the morning came I could not but look at Skiddaw in a different light. Its timeless shape is no such thing. Nor has it been 'here' for ever. Nor again is this a matter of past history alone, for the movement of the continents of course

135

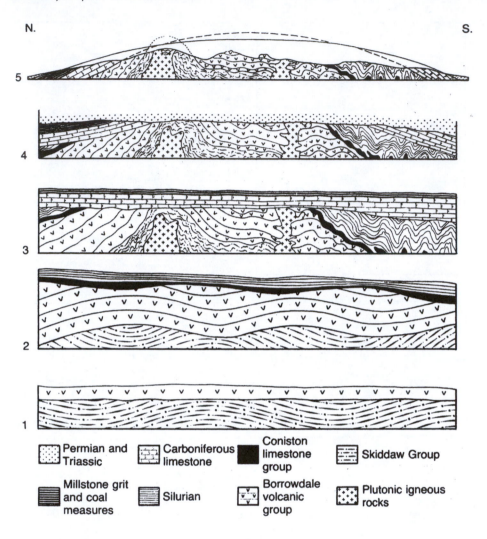

N.   **S.**

Legend:

- ░░ Permian and Triassic
- ▦ Carboniferous limestone
- ■ Coniston limestone group
- ▤ Skiddaw Group
- ▬ Millstone grit and coal measures
- ▤ Silurian
- ⌄ Borrowdale volcanic group
- ▨ Plutonic igneous rocks

1 = deposition of Skiddaw group; folding and erosion; deposition of Borrowdale Volcanic Group
2 = folding and erosion; deposition of Coniston Limestone Group and Silurian rocks
3 = severe folding and great erosion; intrusion of plutonic igneous rocks; deposition of Carboniferous rocks
4 = gentle folding and considerable erosion; deposition of Permian and Triassic rocks
5 = gentle uplift, producing an elongated dome and resulting in radial drainage; erosion to present form

figure 12.4   *The travails en route. Diagrammatic sections to illustrate the building of the Lake District (after Taylor et al., 1971).*

Source: Goudie, A. (1990)

continues (the present is not some kind of achieved terminus) – on average they drift a few centimetres a year: about the rate at which our finger nails grow. And the whole of north west Britain is still rising in relief after the removal of the great weight of ice (while the south east tips compensatorily down). Erosion continues apace. In figure 12.1 the space and the time of this place are separated. The geological series shows 'time', but with no indication of the spatial shifts involved. The geological sketch map, as a classic map, shows a surface as given, but with no indication of the fact that this is a conjunction in movement.

Immigrant rocks: the rocks of Skiddaw are immigrant rocks, just passing through here, like my sister and me only rather more slowly, and changing all the while. Places as heterogeneous associations. If we can't go 'back' home, in the sense that it will have moved on from where we left it, then no more, and in the same sense, can we, on a weekend in the country, go back to nature. It too is moving on.

'Nature', and the 'natural landscape', are classic foundations for the appreciation of place. That literature is too extensive to be addressed here but it does raise important issues. Arif Dirlik (2001) has written thoughtfully about the connection, arguing that 'place is the location … where the social and the natural meet' (p. 18). For him one of the significant implications of this is that it lends place a fixity. Responding, sympathetically, to my own conceptualisation of place, and to those of others, he none the less argues that it can be 'overly zealous, I think, in disassociating place from fixed location. This is where ecological conceptions of place, which are almost totally absent from these discussions (and marginalized by them in the preoccupation with the "social construction of space"), have some crucial insights to contribute by once again bringing nature … into the conceptualization of place' (p. 22). The point about the exclusive focus on human social construction is well taken, and coincides with my intention here. However, Dirlik's reason for bringing nature back in is to emphasise 'the fixity of places' (p. 22), to provide a foundation. And even while he argues that this 'is not the same thing as immutable fixity' (p. 22), the emphasis *is* none the less on fixity. There is again a serious point here – the vast differences in the temporalities of these heterogeneous trajectories which come together in place are crucial in the dynamics and the appreciation of places. But in the end there is no ground, in the sense of a stable position, and to assume there is to fall into those imaginations criticised in Chapter 9 for celebrating a mobile culture while holding (or trying to hold) nature still.

The event of place

And yet, if everything is moving where is here?

Nor, of course, is it just humans and continents that are on the move. Sarah Whatmore has written of the 'mobile lives' of animals and plants – 'on scales that vary from the Lilliputian travels of a dung beetle to the global navigations of migrating whales and birds, … [of] plant seeds journeying in the bellies of animals' (1999, p. 33; see also Clark, 2002; Deleuze and Guattari, 1987). The Lake District has been repopulated, through the movements of animals, plants and humans, in the few thousand years since the last ice age. (So what is indigenous here?) Arctic terns migrate each year between the polar regions; the swifts which nest each year in my road in Kilburn (arriving some time between May Day and the Cup Final) are now as I write this (in January in London) over 7000 miles away in Southern Africa. And the long evolution of patterns of bird migration has been influenced by the drifting of the continents and by the periodic advance and retreat of the succession of ice ages (Elphick, 1995). It is common now to understand 'earth and life' as changing and evolving in relation to each other (see Open University, 1997), to challenge in some way the causal separation of biology and geology. That the organic can affect the tectonic, and so forth. Barbara Bender (personal communication) reflects, when considering Lesternick in south west England, that 'Landscapes refuse to be disciplined. They make a mockery of the oppositions that we create between time (History) and space (Geography), or between nature (Science) and culture (Social Anthropology)'. 'History is no longer simply the history of people, it becomes the history of natural things as well' (Latour, 1993, p. 82). Reading Bruno Latour hints at how social scientists can dispense with our awe of natural science's 'truth' while still (perhaps even in consequence) integrating Skiddaw and weekend tourism as histories/trajectories whose co-formation participates in the event of Keswick. As the train cuts through the chalk hills (the chalk laid down about 100 million years ago and somewhat to the south – see figure 12.3) on the way from London to Milton Keynes it is a tiny thing on a planet spinning on its axis and circling the sun. This corner of the country sinking back down over the millennia since the last ice age. And bouncing gently a couple of times a day, as the tide goes in and out. Cornwall to the west goes up and down by 10 centimetres with each tide. There is no stable point.

> *There are tides in the solid earth as well as in the ocean – every day, for example, the interior of the North American continent goes up and down by about 20 cm. (Open University, 1997, vol. 1, p. 78)*

The various poles have wandered too, and have flipped between each other. Polaris is the northern pole star now, but it was not so when the pyramids were built, between four and five thousand years ago. (I know we all 'know' this; the point is to feel it, to live in its imagination.) Just relative movement.

> *The swifts which leave Kilburn in August do a round trip of up to 15,000 miles, and most of them do not land even once during the 9 months they are away.*

If there are no fixed points then where is here? A thing we now call Skiddaw (even the naming won't stay still, Macpherson as recently as 1901 referring to it as 'Skiddaw (or Skidda)', p. 2) slowly (from my point of view) taking form, still rising, still being worn down (and the constant tramp of hiking boots, not to mention mountain bikes, is a significant form of erosion in the Lake District), still moving on; my sister and I just here for a long weekend, but being changed by that fact too. 'All the essences become events'; place as 'Real as Nature, narrated as Discourse, collective as Society, existential as Being' (Latour, 1993, pp. 82, 90). And space and time, together, the outcome of this multiple becoming. Then 'here' is no more (and no less) than our encounter, and what is made of it. It is, irretrievably, here *and* now. It won't be the same 'here' when it is no longer now.

> *There is 'a consensus that the angle of tilt [of the Earth's axis] has changed significantly over geological time, but in a some-what chaotic manner'. (Open University, 1997, vol. 1, p. 80)*

'Here' is where spatial narratives meet up or form configurations, conjunctures of trajectories which have their own temporalities (so 'now' is as problematical as 'here'). But where the successions of meetings, the accumulation of weavings and encounters build up a history. It's the returns (mine, the swifts') and the very differentiation of temporalities that lend continuity. But the returns are always to a place that has moved on, the layers of our meeting intersecting and affecting each other; weaving a process of space-time.[16] Layers as accretions of meetings. Thus something which might be called there and then is implicated in the here and now. 'Here' is an intertwining of histories in which the spatiality of those histories (their then as well as their here) is inescapably entangled. The interconnections themselves are part of the construction of identity. What Gupta and Ferguson (1992) call 'a shared historical process that differentiates the world as it connects it'.[17]

Source: © Peter Pedley Postcards

I must insist here, quite passionately, on one thing. This is not, as it is on occasions understood to be, a position which is hostile to place or working only for its dissolution into a wider space. Nor is it a deconstructive move, merely exposing an incoherence within an imagined essence (nor indeed is it proposing that what is at issue is purely within the discursive). It is an alternative positive understanding (DeLanda, 2002). This is certainly not to argue against 'the distinctiveness of the place-based' nor – and most particularly – is it to declare 'that there is nothing special about place after all' (Dirlik, 2001, pp. 21 and 22). Quite to the contrary: but what is special about place is not some romance of a pre-given collective identity or of the eternity of the hills. Rather, what is special about place is precisely that throwntogetherness, the unavoidable challenge of negotiating a here-and-now (itself drawing on a history and a geography of thens and theres); and a negotiation which must take place within and between both human and nonhuman. This in no way denies a sense of wonder: what could be more stirring than walking the high fells in the knowledge of the history and the geography that has made them here today.

This is the event of place. It is not just that old industries will die, that new ones may take their place. Not just that the hill farmers round here may one day abandon their long struggle, nor that that lovely old greengrocers is now all turned into a boutique selling tourist bric-à-brac. Nor, evidently, that my sister and I and a hundred other tourists soon must leave. It is also that the hills are rising, the landscape is being eroded and deposited; the climate is shifting; the very rocks themselves continue to move on. The elements of this 'place' will be, at different times and speeds, again dispersed.

(And yet, in its temporary constellation we (must) make something of it.)

This is the event of place in part in the simple sense of the coming together of the previously unrelated, a constellation of processes rather than a thing. This is place as open and as internally multiple. Not capturable as a slice through time in the sense of an essential section. Not intrinsically coherent. As Low and Barnett (2000) argue, many concepts of place are underwritten by 'a notion of uniform time' such that places are conceived of 'as sites where a host of different social processes are gathered up into an intelligible whole' (p. 58).[18] It is an assumption of coherence which is buttressed by that modernist imagination of space as always-already territorialised which was discussed in Chapter 8. To guard against the presumption of coherence (the assumption that all these different constituent processes will somehow coordinate), they argue for working with the term 'conjuncture'. '"Thinking conjuncturally" suggests a shuttling back and forth between different temporal frames or scales to capture the distinctive character of processes which appear to inhabit the "same" moment in time' (p. 59; see, for one attempt at a working through of this in the context of place-definition, Allen et al., 1998). Likewise Dodgshon (1999) writes of 'the false synchronicity of the "moment in being", its deceptive flatness' (p. 615). Nor is this a de-structuring (except – which is post-structuralism's point – to some existing imaginations). It is simply a coming together of trajectories.

But it is a uniqueness, and a locus of the generation of new trajectories and new configurations. Attempts to write about the uniqueness of place have sometimes been castigated for depoliticisation. Uniqueness meant that one could not reach for the eternal rules. But 'politics' in part precisely lies in not being able to reach for that kind of rule; a world which demands the ethics and the responsibility of facing up to the event; where the situation is unprecedented and the future is open. Place is an event in that sense too.

Reconceptualising place in this way puts on the agenda a different set of political questions. There can be no assumption of pre-given coherence, or of community or collective identity. Rather the throwntogetherness of place demands negotiation. In sharp contrast to the view of place as settled and pre-given, with a coherence only to be disturbed by 'external' forces, places as presented here in a sense necessitate invention; they pose a challenge. They implicate us, perforce, in the lives of human others, and in our relations with nonhumans they ask how we shall respond to our temporary meeting-up with these particular rocks and stones and trees. They require that, in one way or another, we confront the challenge of the negotiation of multiplicity. The sheer

fact of having to get on together; the fact that you cannot (even should you want to, and this itself should in no way be presumed) 'purify' spaces/places. In this throwntogetherness what are at issue are the terms of engagement of those trajectories (both 'social' and 'natural'), those stories-so-far, within (and not only within) that conjuncturality. As Donald (1999) writes in his more specific consideration of cities as places, politics is the (ever-contested) question of our being-together. This is one part of the 'responsibilities' of place to which Part *Five* will turn.

(Geographies of knowledge production 2: places of knowledge production)

'Science parks' are among the most potent icons of the knowledge economy which, we are constantly told, characterises today's, and tomorrow's, global capitalism. They are among the carefully chosen and designed sites of the production of an electronically connected world (Chapter 9). They are also one element in an emerging, violently unequal, twenty-first-century geography of, a particular form of, knowledge. Demarcated, landscaped enclosures dedicated to the production of science (usually, specifically, commercialisable science), these are 'places' of a kind; constructed places, coherent, planned (ironic, isn't it, in this soi-disant age of anti-plan).

Easily recognisable, replicated over and over, they are scattered around the planet like flags on a map, each witness to some local/regional/national desperation to create another Silicon Valley, jump-start another Cambridge Science Park, or at least attract a few bits of 'high technology'. The requirements, to be able to play this industrial location game, are: an enclosed and separate space; a landscaped environment within, to give off some evocation of 'quality'; a publicity blurb which emphasises the nearby university (as elite-sounding as possible); and a picturing of the wider environmentally attractive area within which it is set (where 'environmentally attractive' stands for a very particular aesthetic favouring a tamed suburban 'rurality', and a definite absence of the ruins of nineteenth/twentieth-century industrialisation). Preferably, since these knowledge-intensive sectors have a tendency to cluster, you need also to be able to demonstrate to potential investors that others like them have already made this choice (they would not want to be pioneers, or take a risk). These are some of the 'location factors' you will need to parade in order to attract this part of the new knowledge economy (Massey et al., 1992).

All this is well known, and some of the contradictions of it are immediately evident. The knife-sharp class-ridden nature of it all, and the inevitably greater success in areas precisely not 'marred' by the decline of previous eras, mean that these agents of economic regeneration produce 'regeneration' precisely where it is least needed. And so on.[19]

There is another way of reading these constructed places. Entangled and enfolded within them is a multiplicity of trajectories each of which has its own spatiality and temporality; each of which has been, and still is, contested; each of which might have turned out quite differently (yet where the intersection of these histories has often served to reinforce the existing lines of dominance).

The particular form of the proliferation of the division of labour within industry which resulted in that (so well known it seems natural) separation of 'conception' *from 'execution' was propelled by forces both of class and of a particular notion of knowledge. Knowledge as removable from the shop floor, for instance. Knowledge as separable rather than tacit; distanced rather than embedded and embodied. It resonates with the abstractions discussed in Part* Three: *'the way in which a science, or a conception of science, participates in the organization of the social field, and in particular induces a division of labor, is part of that science itself' (Deleuze and Guattari, 1988, pp. 368–9). The separation and the class nature of this division of labour were sharply reinforced by geographical division and distance: a dispersion of industrial sites emerged, with clearly distinct characteristics (a particular spatial division of labour), the spatiality being integral to the proliferation of divisions among the workers and the reinforcing of their differentiated characteristics.[20] It is a recapitulation of an old story in Western history: the spatial seclusion of the desert for early Christian thinkers, the emergence of monasteries as elite places of knowledge production, the mediaeval universities. All of them places which crystallise through spatialisation a separation of Mind from Body, a notion of science as removal from the world. A material spatialisation of Stengers' account of science's dismissal of mere phenomena, and of Fabian's tale of the distancing of knowing subject from object of knowledge. Here in the places of high technology these structurings of the knowledge relation are deeply interwoven with those of class, and the two together are reinforced through spatial form.*

That is one strand of the spatial histories these places enfold. Another is that, through Western history, they have been part and parcel of the struggle around the creation of intelligible genders, of certain forms of 'masculine' and 'feminine'. Over and over again the establishment of these places was bound up with the distinction of genders and the expulsion of women. Brown, writing of one of the earliest of such spaces, tells that 'Fear of women fell like a bar of shadow across the paths that led back from the desert into the towns and villages' (1989, p. 242), and David Noble in his wonderful account of this winding history over two millennia, writes of 'the male monastic flight from women' (1992, p. 77) and documents in detail the embattled continuation of this flight into the university and into modern science.[21] (One is drawn to reflect on the postmodern return to the desert, or at least to the figure of the desert – the space of an absence of women?) A long history, in fact, not just of the exclusion of women but of the contested constitution of what it was going to mean to be a (certain kind of) man or woman. The 'masculinity' of the world's science parks today is not just a product of, nor can it be measured by, the fact of the overwhelming dominance on them of male employees. It is an outcome of a longer deeper history of gender construction which itself was/is spatially embedded within the making of defensive, specialised, 'places of knowledge'.

And finally (for our purposes here) a third trajectory: these places of knowledge production were all also elite places of the production of legitimate, recognised, authorised knowledge. For there were always, and are still, other forms of knowledge: in the society that lay beyond the walls, in the villages along the edges of the desert, on the shop floor of the places of material production banished to the geographical 'periphery'. The time-spaces of mediaeval monasteries, the old universities and today's science parks are all of them moments in the interweaving of the histories of the legitimation of a certain form of knowledge production, the generation and maintenance of a masculinised caste which specialises in the definition and production of that knowledge, and the moulding of that kind of masculinity itself.

These trajectories together have propelled the exclusions on which science parks have been constituted. They are, moreover, interwoven histories each of which has been contested. In that sense these spaces are both an achievement and still open to challenge (see Chapter 5). Noble (1992) recounts in detail the battle over gender, and the struggle to maintain an authorised elite can be traced from the battles within early Christianity, through Paracelsus, through the riot of dissidence over centuries in Europe (Lollards, Anabaptists, Muggletonians, early Swedenborgians, Brownists, Baptists, Quakers, Ranters ...) to the Lucas Aerospace workers of the final decades of the twentieth century.[22] The times of these places are many. Science parks embody not only recent economic calculation but also long histories of social struggle, over the nature and ownership of knowledge, over the meanings and delineations of gender, over the material establishment in lived relations of the philosophical postulation of an opposition of Mind and Body. These things are built into the very fabric of such places as the physical and social precipitates of particular intersections of a multiplicity of trajectories. And, in spite of their neatly manicured appearance, the histories they embody do not coalesce into a simple coherence. The contests in the histories they embody erupt at different moments, dislocating in different ways.

These are particular, and particularly powerful, spatial formations. They articulate in physical form both the social spatiality of knowledge production and an imagined spatiality of the knowledge relation. It is a longer and more multiple story than the one told by Stengers; one in which the choice between Einstein and Kepler was but an episode; and it is a history in which geography was crucial.

These, then, again, are places as temporary constellations where the repercussions of a multiplicity of histories have been woven together. Knowledge production and legitimation function here as practices which generate space-times (as well as concepts of space-time). Place as event. Ironically, these high-tech places are controlled and planned events. Their components are disciplined, down to the enrolment of the non-human, in suitable, domesticated forms ('tasteful' landscape, watered lawns), to bolster their cachet. 'Ironically' because these 'places of innovation' seem designed to limit their potential character as places as innovation. And yet of course, in the end, the potential event of place remains. The containment is impossible.

Hamburgs ältester Einwanderer!

Neues Staatsangehörigkeitsrecht
"Partnerschaft für Integration" eine Initiative der Ausländerbeauftragten

Informationen beim Einwohner-Zentralamt, Staatsangehörigkeits- und Einbürgerungsangelegenheiten sowie den deutsch-ausländischen Begegnungsstätten und den Sozialberatungsstellen der Wohlfahrtsverbände oder im Internet: www.Hamburg.de/Behoerden/Auslaenderbeauftragter/welcome.htm

Part *Five*
A relational politics of the spatial

In Bruno Latour's political proposal for 'A (philosophical) platform for a left (European) party' (1999a), the third of his ten planks begins 'I have the feeling that we are slowly shifting from an obsession with time to an obsession with space' (p. 14), and a little further on he reflects that 'If, as philosophers argue, time is defined as the "series of succession" and space as the "series of simultaneity", or what coexists together at one instant, we might be leaving the time of time – successions and revolutions – and entering a very different time/space, that of coexistence' (p. 15). I have reservations about this formulation. It itself, somewhat contradictorily, has the flavour of linear temporality and singular movement; its account of the emergence of the spatial relies on the temporal in precisely the way that Grossberg criticises (see Part *Two*); and I am not sure whether, in fact, such a shift is occurring. Certainly, too, I would not want to argue for an obsession with space, nor the replacement of time by space; nor am I simply dismissive of all previous politics of the left.

And yet I do want to argue, in tune with Latour's vision, for a politics, perhaps better an angle of vision *on* politics, which can open itself up in this way to an appreciation of the spatial and the engagements it challenges us to. That is to say, less a politics dominated by a framing imagination of linear progression (and certainly not singular linear progression), and more a politics of the negotiation of relations, configurations; one which lays an emphasis on those elements addressed in Chapter 10: practices of relationality, a recognition of implication, and a modesty of judgement in the face of the inevitability of specificity.

Latour writes of 'the new obligations of coexistence (that is the production of space), of heterogeneous entities no-one can either simplify or eliminate for good' (p. 15). Again, the term coexistence is perhaps inadequate: stress needs to be laid also on coformation, and on the inevitability of conflict. What is at issue is the constant and conflictual process of the constitution of the social, both human and nonhuman. Such a view does not eliminate an impetus to forward movement, but

it does enrich it with a recognition that that movement be itself produced *through* attention to configurations; it is out of them that new heterogeneities, and new configurations, will be conjured. This is a temporality which is not linear, nor singular, nor pregiven; but it is integral to the spatial. It is a politics which pays attention to the fact that entities and identities (be they places, or political constituencies, or mountains) are collectively produced through practices which form relations; and it is on those practices and relations that politics must be focused. But this also means insisting on space *as* the sphere of relations, of contemporaneous multiplicity, and as always under construction. It means not falling back into those strategies of evasion which fail to face up full on to the challenge of space.

This is a change in the angle of vision away from a modernist version (one temporality, no space) but not towards a postmodern one (all space, no time) (see Chapter 7); rather towards the entanglements and configurations of multiple trajectories, multiple histories. Moreover, what this means in turn is that the politics itself might require a different geography: one which reflects the geography of those relations. This Part attends to some of those geographies: to negotiations within place, to the challenge of linking local struggles, to the possibility of an outwardlooking local politics which reaches out beyond place.

13
throwntogetherness: the politics of the event of place

In the autumn of 1999 workers labouring on the bed of the river Elbe where it begins to open out to the sea at Hamburg came up against a massive boulder. It was a noteworthy event and made the news. The rock became popular and the people of Hamburg began to visit it. But this celebrated resident of the city turned out to be an immigrant. It is an erratic, pushed south by the ice thousands of years ago and left here as the ice retreated. By no means, then, a 'local' boulder.

Or is it? How long do you have to have been here to be local?

On 1 January 2000, German citizenship laws were relaxed somewhat and Ulla Neumann, the imaginative official for foreign immigrants in Hamburg, seized upon the immigrant boulder and the practices it had engendered; to raise questions, to urge a reimagining of the city as open, with the aim of its being lived more openly. The poster in figure 13.1, designed by Steffan Böhle, was the result. Some established immigrants were to be granted citizenship, to be accepted – like the rock – as 'of the place'. The design of the poster reinforced the argument. Hamburg as a major port and very visibly open to ships and workers and capital from around the world had long evoked one image of the city as cosmopolitan. There was an established and much-used logo: 'Hamburg: gateway to the world'. The poster, with the gateway cut through the immigrant rock, and with the city visible through it, both addressed a challenge to established German citizens to make this logo (this already-existing self-image) meaningful in another way, to take it at its word and press it home, and offered an invitation to immigrants to find out more.[1]

It was an attempt to urge an understanding of this place as permeable, to provoke a living of place as a constellation of trajectories, both 'natural' and 'cultural', where if even the rocks are on the move the question must be posed as to what can be claimed as belonging; where, at the least, the question of belonging needs to be framed in a new way. The gateway through the rock speaks of openness and migrants and lays down the challenge of the possibility of living together.

The poster plays to the way in which people live the city, practise it in a whole variety of ways, as they constantly make space-place. It is intended to be

Hamburgs ältester Einwanderer!

Neues Staatsangehörigkeitsrecht
"Partnerschaft für Integration" eine Initiative der Ausländerbeauftragten

Informationen beim Einwohner-Zentralamt, Staatsangehörigkeits- und Einbürgerungsangelegenheiten sowie den deutsch-ausländischen Begegnungsstätten und den Sozialberatungsstellen der Wohlfahrtsverbände oder im Internet: www.Hamburg.de/Behoerden/Auslaenderbeauftragter/welcome.htm

figure 13.1   *'Hamburg's Oldest Immigrant'*

Source: Design © Steffan Böhle; used with the kind permission of
      Ulla Neumann

an active agent in that refiguring, reconstituting Hamburgers' story of their past in order to provoke a reimagination of the nature of the present. Its intent is to mobilise a political cosmology, in Fabian's (1983) terms, but a political cosmology which does not somehow exist prior to but is part and parcel of the way in which we live and produce time-space. As Ingold writes, 'the forms people build, whether in the imagination or on the ground, arise within the current of their involved activity, in the specific relational contexts of their practical engagements with their surroundings' (1995, p. 76). A knowledge of the city produced through engagement. We Hamburgers love that boulder, we have accepted it into the city; an important element in our practised relation to the city, indeed one of its iconic emblems, is a migrant.[2] An already instituted practice might shift our imagination which might provoke a reconsideration of (or at least more debate about) other practices.

Place as an ever-shifting constellation of trajectories poses the question of our throwntogetherness. This is Kevin Robins' point in insisting on the importance of material place (Chapter 9). The chance of space may set us down next to the unexpected neighbour. The multiplicity and the chance of space here in the constitution of place provide (an element of) that inevitable contingency which underlies the necessity for the institution of the social and which, at a moment of antagonism, is revealed in particular fractures which pose the question of the political. James Donald (1999), wrestling with the nature of the social and the political in the city, writes that 'We experience our social world as simply the way things are, as objective presence, because that contingency is systematically forgotten' (p. 168). Drawing on Laclau, he argues that, although we cannot hope to capture the fullness of that contingency, it does at particular moments present itself before us.[3] It is the undecidability of the essential contingency which makes possible the opening up of the field of the political: 'The moment of antagonism where the undecidable nature of the alternatives and their resolution through power relations becomes fully visible constitutes the field of the "political"' (Laclau, 1990, p. 35; cited in Donald, 1999, p. 168). *Hamburgs ältester Einwanderer!*, the poster, places itself at that moment, unsettling the givenness.

Places pose in particular form the question of our living together. And this question, as Donald also argues, through reference to Mouffe (1991), Nancy (1991) and Rajchman (1991, 1998), is the central question of the political. The combination of order and chance, intrinsic to space and here encapsulated in material place, is crucial. 'Chaos is at once a risk and a chance', wrote Derrida (1996). And Laclau argues that the element of dislocation opens up the very possibility of politics. Sennett (1970) urges us to make use of disorder, and Levin (1989) evokes 'productive incoherence'. The passage from Derrida runs like this:

> This chaos and instability, which is fundamental, founding and irreducible, is at once naturally the worst against which we struggle with laws, rules, conventions, politics and provisional hegemony, but at the same time it is a chance, a chance to change, to destabilize. If there were continual stability, there would be no need for politics, and it is to the extent that stability is not natural, essential or substantial, that politics exists and ethics is possible. Chaos is at once a risk and a chance. (p. 84)

The relation to spatiality is two-fold: *first* that this irreducibility of instability is linked to, and certainly conditional upon, space/spatiality and *second* that much 'spatial politics' is concerned with how such chaos can be ordered, how juxtapositions may be regulated, how space might be coded, how the terms of connectivity might be negotiated. Just as so many of our accustomed ways of imagining space have been attempts to tame it.

The space we call 'public space' raises these arguments most pointedly. There is widespread concern about 'the decline of public space' in the neoliberal city: the commercial privatisation of space, the advent of new enclosures such as, iconically, the shopping mall, and so forth. These are clearly processes we may witness with alarm, and for a number of good reasons. They involve the vesting of control over spaces in the hands of non-democratically-elected owners; they may involve the exclusion from many such spaces of groups whom we might have expected (for instance had the space been publicly owned) to have been allowed there (the exclusion of unemployed 'loiterers' – deemed not to be prospective shoppers – from shopping malls has probably emerged as the most-cited example). These are serious issues. But the tendency to romanticise public space as an emptiness which enables free and equal speech does not take on board the need to theorise space and place as the product of social relations which are most likely conflicting and unequal. Richard Rogers' call, in his report *Towards an urban renaissance* (Urban Task Force, 1999), for more public spaces in the city envisages them as squares, piazzas, unproblematically open to all. While one might share his desire for a greater presence of this element of the urban fabric, its 'public' nature needs to be held up to a scrutiny which is rarely devoted to it. From the greatest public square to the smallest public park these places are a product of, and internally dislocated by, heterogeneous and sometimes conflicting social identities/relations. Bea Campbell's ('public') shopping centres in *Goliath* (1993) dominated by different groups at different times of day and night (and dominated in explicitly excluding ways) are a good example (Massey, 1996b). In London there has been the sharpest of spats over the presence of pigeons (a tourist attraction, beloved by all, animals with rights *versus* pigeons as a flying, feathered health hazard) in Trafalgar Square. *Comedia*'s (1995) study of public parks pointed clearly to the continuing daily negotiations and struggles, sometimes quiet and persistent, sometimes more forceful, through which day in day out these spaces are produced. Such 'public' space, unregulated, leaves a heterogeneous urban population to work out for itself who really is going to have the right to be there. All spaces are socially regulated in some way, if not by explicit rules (no ball games, no loitering) then by the potentially more competitive (more market-like?) regulation which exists in the absence of explicit (collective? public? democratic? autocratic?) controls. 'Open space', in that particular sense, is a dubious concept. As well as objecting to the new privatisations and exclusions, we might address the question of the social relations which could construct any new, and better, notion of public space. And that might include, sometimes, facing up to the necessities of negotiated exclusion.

There is a further point. Rogers reflects Walzer (1995) in working with a notion of open-minded spaces. But this must be seen as an asymptotic process. There may be parallels here with Derrida and with theorists of radical democracy and notions of democracy-to-come, of a continually receding horizon of the open-minded-space-to-come, which will not ever be reached but must constantly be worked towards. This is like Robbins' 'phantom public sphere': a fantasy, but one which it is imperative that we continue to pursue. In Rosalyn Deutsche's words, 'If "the dissolution of the markers of certainty" calls us into public space, then public space is crucial to democracy not despite but because it is a phantom' (1996, p. 324). By the same token, and precisely because of the elements of chaos, openness and uncertainty which they both embody, space, and here specifically place, are potentially creative crucibles for the democratic sphere. The challenge is having the confidence to treat them in this way. For instituting democratic public spaces (and indeed the spaces of places more generally) necessitates operating with a concept of spatiality which keeps always under scrutiny the play of the social relations which construct them. 'Instead of trying to erase the traces of power and exclusion, democratic politics requires that they be brought to the fore, making them visible so that they can enter the terrain of contestation' (Mouffe, 1993, p. 149).

The argument is not that these places are not public. The very fact that they are necessarily negotiated, sometimes riven with antagonism, always contoured through the playing out of unequal social relations, is what renders them *genuinely* public. Deutsche, in her exploration of the possible meaning of public art, draws on Claude Lefort: 'The hallmark of democracy, says Lefort, is the disappearance of certainty about the foundations of social life' (p. 272). 'The public space, in Lefort's account, is the social space where, in the absence of a foundation, the meaning and unity of the social is negotiated – at once constituted and put at risk. What is recognised in public space is the legitimacy of debate about what is legitimate and what is illegitimate' (p. 273). As Deutsche reflects, 'Conflict is not something that befalls an originally, or potentially, harmonious urban space. Urban space is the product of conflict' (p. 278).

What applies to public space applies *a fortiori* to more ordinary places. These temporary constellations of trajectories, these events which are places, require negotiation. Ash Amin (2002) writes of such a politics of place as suggesting a different vocabulary: one of local accommodation, a vocabulary which addresses rights of presence and confronts the fact of difference. It would be a vocabulary irreducible to a politics of community and it articulates a politics without guarantees. Moreover, places vary, and so does the nature of the internal negotiation that they call forth. 'Negotiation' here stands for the range of means through which accommodation, anyway always provisional, may be reached or not.

Chantal Mouffe defines the political as being predicated upon 'the always-to-be-achieved construction of a bounded yet heterogeneous, unstable and necessarily antagonistic "we"' (quoted in Donald, 1999, p. 100). Some kinds of places, on certain occasions, do require the construction of such a 'we', but most 'places' in most quotidian ways are of a much vaguer sort. They do not require the constitution of a single hegemonic 'we' (though there may be a multiplicity of implicit ones being wielded in the daily practices that make the place).[4] Jean-Luc Nancy offers the notion of the political as 'a community consciously undergoing the experience of its sharing' (1991, p. 40). The daily negotiation and contestation of a place does not require in quite that sense the conscious collective contestation of its identity (however temporarily established) nor are there the mechanisms for it. But insofar as they 'work' at all places are still not-inconsiderable collective achievements. They are formed through a myriad of practices of quotidian negotiation and contestation; practices, moreover, through which the constituent 'identities' are also themselves continually moulded. Place, in other words does – as many argue – change us, not through some visceral belonging (some barely changing rootedness, as so many would have it) but through the *practising* of place, the negotiation of intersecting trajectories; place as an arena where negotiation is forced upon us. The terms on which it takes place may be the indifference of Young's unassimilated otherness, or the more conscious full interaction which Sennett seeks, or a more fully politicised antagonism.

Donald cites Derrida's *Politics of friendship* on the distinction between respect and responsibility. It is a distinction Derrida aligns with his interpretation of the difference between space and time. Respect, he says, refers to distance, to space, to the gaze; while responsibility refers to time, to the voice and to listening (see Donald, 1999, p. 166). Derrida writes: 'There is no respect … without the vision and distance of a *spacing*. No responsibility without response, without what speaking and hearing *invisibly* say to the ear, and which takes *time*' (1997, p. 60; emphasis in the original, cited in Donald, 1999, p. 166). One might be wary of elements in this formulation including that particular way of differentiating space and time, though the aspect of space as the social is clear. None the less, what 'places' – of all sorts – pose as a challenge and a responsibility is precisely what Derrida is after, the co-implication of his 'responsibility' and 'respect' – might one say time-space? – the recognition of the coevalness (and in 'place' co-presence) of a multiplicity of trajectories.

'Place' here could stand for the general condition of our being together (though it is meant here more specifically than that). However, the spatiality of the social is implicated at a deeper level too. First, as a formal principle it is the spatial within time-space, and at this point most specifically its aspect of being the sphere of multiplicity, and the mutual opacity which that necessarily entails, which requires the constitution of the social and the political. Second, in political practice much of this constitution is articulated through the negoti-ation of places in the widest sense. Imaginations of space and place are both an element of and a stake *in* those negotiations. Hamburg's poster catches precisely at this.

This view of place is most often evoked when discussion turns to that metropolitan-academic preoccupation: *cities*. Donald's careful and stimulating discussion concerns cities specifically. He cites the inevitability of conflict in cities; the challenge of living together in such space–places (that the important question is less the one so often posed – how do I live in the city – but how do we live together – p. 139); he cites Rajchman's question of being 'at home' in a '"world where our identity is not given, our being-together in question." That is the specific sense in which city life is inescapably political' (1999, p. 155). Cities are perhaps the places which are the greatest challenges to democracy (Amin et al., 2000). They are peculiarly large, intense and heterogeneous con-stellations of trajectories, demanding of complex negotiation.[5] This imagination of the (usually Western) city, however, has most often focused on cultural and ethnic mix – which is certainly one kind of meeting of trajectories effected through neoliberal globalisation. But there are other ways, too, in which such cities, and perhaps especially Western so-called 'world cities', have been the site of the colliding trajectories of globalisation.

Take London. London is a world city for capital as well as for international migration. The trajectories of capital, just as much as of ethnicity, come into col-lison here. Trading on its long history as mercantile hub of empire, London has gathered into itself a huge constellation of financial and associated functions. The financial City marks the city (the impossibility of distinguishing between them in speech provokes wandering Derridean thoughts). The City's trajectory is massive and (even allowing for acknowledged weaknesses and vulnerabili-ties) forceful. It is also a trajectory which is outwardlooking; its gaze sweeps the planet. Until the recent opening up of 'property-development opportunities' there, the City knew more about markets on distant continents than about what was happening just across the river. Moreover this is a trajectory which collides here in London with other economic histories which have, so far, continued to be made in this place. There are the remains of physical trade, a million service industries, national, local and international, a considerable manufacturing base and a tattered public sector infrastructure. These are trajectories with different resources, distinct dynamics (and strengths in the market) and temporalities, which have their own directions in space-time, and which are quite differently embedded within 'globalisation'.

It is a real collision. The dominance of London by global financial industries changes the character and the conditions of existence of all else.[6] The working of this collision through land prices is the most evident of these effects. Manufacturing industry which might otherwise have survived is made uneconomical by the price it has to pay for land/premises. The continuing profitability of the process of production, before such costs are taken into account, is nullified by the inability to find or retain a site in the face of the voracious demand and the greater ability to pay, on the part of these 'world city' industries. Put another way, the growth of the City is an element in the production of unemployment among manufacturing workers. It places constraints on and presents obstacles to the growth, sometimes even the survival, of other parts of London's economy. Infrastructure is straining at the seams, its efficiency declining, and capacity problems are evident everywhere. The grotesquely high wages in the City have further knock-on effects, on prices in general but on housing costs in particular. It becomes impossible to sustain a public sector because public sector workers (given central government policy) cannot afford to live here. Even in my own neck of the woods, on the other side of London from the City, a 'local community policeman' has to commute in from Leicester; and a letter was dropped through my door (and through all the letterboxes in the area) interpellating me, and the rest of this area, through a specific bit of our identity (to 'The Home Owner' it said): and it went on to invite me to take advantage of the fact that I live in the same metropolis as the overpaid cohorts of global finance. Their annual bonuses would be pushing up house prices – maybe I wanted to sell.

This, then, is a clash of trajectories where the dominance of one of them reverberates through the whole of London: changing the conditions for other industries, undermining the public sector, producing a greater degree of economic inequality in London than in any other city in the UK (and that last fact in itself has effects on the lives of everyone). London's higher 'average' salaries conceal a vast inequality – but the additional costs which the high end of that distribution produces have to be borne by everyone.

London is a 'successful' city. Endlessly it is so characterised. (The other regions of the country are problems, we are told, but not London and the South East.) Yet the same documents almost invariably then go on to hint at a difficulty with this characterisation. London is a successful city, they aver, 'but there are still great areas of poverty and exclusion'. Spokespeople for London point to this evident fact in claims for a greater share of the national cake. Prime Minister Tony Blair deploys it constantly in his attempt to evade the issue of inequality between regions (there's poverty in London, too, you know …). (What is needed, of course, is redistribution *within London* – see Amin et al., 2003.)

The problem is in the conjunction. First in the conjunction 'but'. The sentence should rather read: 'London is a successful city *and partly as a result of the terms of that success* there are still great areas of poverty and exclusion.' And second, in the conjunction of trajectories of the economy: the huge concentration of world city industries (and especially finance) is one element in the constellation of forces *producing* that poverty and exclusion.[7]

This is a material collision, moreover, which forces political choice. What is to be the economic strategy of the city? At present it is simply to prioritise finance as the key to world citydom. But the fact that London's 'success' is one of the dynamics producing poverty and exclusion implies at least a query as to the meaning of this word 'successful' and should raise a question about the model of growth. It makes no sense to go on promoting 'growth' in the same old way (not, that is, if the aim, as constantly stated, is to *reduce* poverty and exclusion). Clearly, then, a decision has to be made: between reducing poverty and promoting the City. It is a real political choice. The very suggestion generates anxiety: to take one's foot off the accelerator might mean finance would flee to Frankfurt. This is the reply which is endlessly offered. And who knows how much truth there might be in that fear/threat? The point is that if there is *any* truth in it then there are mutually exclusive (antagonistic) options in front of us: on the one hand policies which favour the City and on the other policies which aim straight at redistribution. This collision of trajectories in place highlights a conflict which requires a political stance.[8]

It is a conflict which is usually hidden. Indeed the real difficulty *is* that lack of recognition. There is a refusal to recognise the antagonism. To those who point to the need to address the problem of poverty the response begins with political agreement. *Of course* they want to address poverty and exclusion (actual redistribution is less easily acceded to). This will be done by multiplier effects from the City (but we know that trickle-down doesn't work); or, a more recent version, soon virtually everyone will be drawn into this new economy (so who, then, will empty the dustbins, nurse the sick, be our local community policeman …?).

At such a point, the argument can become a seemingly technical one over means of achievement. But what has really happened is that the antagonism has been displaced. Rather than an explicit conflict over political aims what we have now is a confrontation between imaginations of the city. The pro-finance view often rests upon a contrast between 'new economy' and 'old', supported by the myth of the new economy as panacea. (The centuries-old financial City is here – ironically – cast as 'new' in opposition to manufacturing as 'old'!) In this imaginary the economy has a classy centrepiece with the rest of the population finding a role in servicing it. It is this structure which produces trickle down and multipliers to all. It is a unity. And it is a unity rhetorically bolstered through recourse to the establishment of external enemies: the other regions of the country (accused of taking too big a share, through redistribution, of the national tax revenue); and Frankfurt (portrayed as forever standing ready to take over as financial capital of Europe). The alternative imaginary refuses this proclaimed unity and instead stresses the multiplicity and interdependence of the various parts of the urban economy, together with recognition of the dislocations, the clashings of diversity, within it. An imagination of a simply coherent entity, with finance as the shining pinnacle, the engine of growth pulling all else along, but with some problems of internal uneven development still to be smoothed out, confronts an imagination of this place as a clash of trajectories of differential strength and where that differential strength is part of what must be negotiated. What is in dispute is what Rajchman has called the 'principle of the spatial dispositions of our being together' (1998, p. 94). Sometimes you have to blow apart the imagination of a space or place to find within it its potential, to reveal the 'disparition' 'in what presents itself as a perceptual totality' (p. 19). To challenge the class politics of London the city itself has to be reimagined as a clash of trajectories.

This itself, however, renders intervention even more tricky. For this has to be an intervention into a constellation of trajectories which, though interacting and undoubtedly affecting each other, have very different rhythms. There is no coherent 'now' to this place (Chapter 12). The thing which is place is not the closed synchrony of structuralism, nor is it the frozen slice-through-time which has so often been characterised as space. All of which has further implications for politics. It means that the negotiations of place take place on the move, between identities which are on the move. It also means, and this is more important to the argument here, that any politics catches trajectories at different points, is attempting to articulate rhythms which pulse at different beats. It is another aspect of the elusiveness of place which renders politics so difficult.

So, in London, progressive people want to solve in the short term the evident need for affordable housing, want larger regional differentials in wage rates (the London Weighting), argue that the 'national' minimum wage ought to be higher in the capital: in other words they want to ameliorate some of the problems posed by the dominance of the City. It is hard not to be sympathetic. Yet such a response will only fan the flames of the longer-term dynamic of the financial world city trajectory. (Yes the financial City can keep growing and somehow we will manage to service it.) Not only is this a patch-and-mend approach to London's economy, not only will such measures through market forces become inadequate almost as soon as they are implemented, but precisely by responding only to immediate processes they perpetuate the long-term dynamics (the dominance of finance, nationally increasing inequality, exacerbating regional uneven development) which lie at the root of it. In the long term such an approach could make things worse (on the redistributors' own criteria).

All this is about cities, and a world city at that. But multiplicity, antagonisms and contrasting temporalities are the stuff of all places. John Rajchman (2001) has reflected upon the current intellectual infatuation (again) with cities: a transdisciplinary obsession. There has, he argues, been a long historical relation between philosophy and the city which has taken the form both of the city providing the conditions for the emergence of philosophy and of philosophy's being the 'city in the process of thinking' (p. 3) – the city as a provocation to philosophy in which 'a city is not only a sociological object, but also a machine that undoes and exceeds sociological definitions posing new problems for thinking and thinkers, images and image-makers' (p. 14). The city as productive of moments of absolute deterritorialisation and, continuing in Deleuzo-Guattarian vein, thus producing too a counterposition between 'the historical deterritorializations of the city' and 'the identities of states and the stories they tell of themselves' (Rajchman, 2001, p. 7) (a contrast which might reflect that between places as simply the unnamed juxtapositions of trajectories which require negotiation, and places with hegemonising identities, with stories 'they' tell of themselves). As Rajchman puts it, Benjamin and Simmel can both be read, in very different ways, as thinkers 'who saw in the peculiar spaces of the metropolis a way to depart from the more official philology or sociology of the German university to explore a zone that could no longer quite be fit[ted] within the great schemes of history and society of the day' (p. 12), an idea which Deleuze would generalise to a philosophy of society as always *en fuite*. It is a wonderfully provocative argument. And it leads Rajchman on to ask what different deterritorialisation is opened up by cities today: what kinds of lines of flight of thought take off 'when we start to depart from ways we have been determined to be towards something other, we are not yet quite sure what …' (p. 17).

Maybe it is indeed that cities have been so productively both condition of and provocation to new thinking. Moreover, part of what this provocation has entailed (though not always explicitly) is a rethinking of city *space* – as accumulation of layers, as ungraspable juxtapositions, and so forth. This space is not, however, unique to the space of the city. It may be the extremity of cities which provokes for some a reimagining, but the in-principle nature of the spatiality is not confined to the urban.

The 'countryside' (such English visions arise, of security and stability) can be deterritorialising of the imagination too. The erratic boulder in Hamburg, the migrant rocks which currently exist as Skiddaw, speak to the same 'new' spatiality as does the city, and open up more widely an appreciation of the temporary nature of the constellation which is place. Tectonic shifts, the ebb and flow of icecaps, the arrival of nonhuman and human migrants; that radical difference in temporalities emphasises more than cities ever can that a 'constellation' is not a coherent 'now'. The persistent focus on cities as the sites which most provoke disturbance in us is perhaps part of what has tamed (indeed is dependent *upon* the taming of) our vision of the rural. Yet reimagining countryside/Nature is more challenging still than responding to the changing spatiality (customarily figured as predominantly human) of the urban.

It is amazing how often this is missed, by even the most self-professedly nomadic of thinkers. Félix Guattari, whose notions of change are otherwise so strong, none the less in his *The three ecologies* (1989/2000) writes of 'natural equilibriums' (p. 66) and, even more bizarrely even if in metaphorical reference to making the desert bloom, of bringing vegetation back to the Sahara (also p. 66). The translator's introduction, too, reinforces this impression of a 'nature' which, if not interfered with by humans, would be 'in balance' (see, for instance, pp. 4 and 5). Or again, Brian Massumi (1992) urges that 'The equilibrium of the physical environment must be reestablished, so that cultures may go on living and learn to live more intensely, at a state far from equilibrium' (p. 141). Such dualisms, as argued in Chapter 9, are inherent in much of the writing of such as Giddens and Beck about 'the risk society'. While cultural mobility and mutability is celebrated, 'disturbances' of nature's pattern are viewed with alarm:

> What seems to underpin the new cosmopolitan environmentalism … is the premise that, left to itself, nature is docile; it maintains its given forms and positions. Culture on the other hand, is seen to be inherently dynamic, both self-transforming and responsible for the mobilization and transmutation of the material world – for better or worse. … Western thought's most pervasive dualism, we might be forgiven for thinking, has returned to haunt cosmopolitan risk society. (Clark, 2002, p. 107)

It is an imagination which fails entirely to appreciate that 'traffic which is nature's own' (p. 104), or to understand the 'indigeneity' of plants and animals, and of rocks and stones, as no less elusive than that of humans.

The nonhuman has its trajectories also and the event of place demands, no less than with the human, a politics of negotiation. It is such a set of negotiations, and maybe in a serious sense frequently failed negotiations given 'nature's' reply, that Mike Davis (2000) documents in his glorious account of Los Angeles. (For the city and nature are not geographically distinct: Whatmore and Hinchliffe, 2002/3.) The production of Los Angeles as it is today, in its conflictual and often perilous throwntogetherness of nonhuman and human, has involved culture clashes (with temperate zone geomorphologists and climatologists misinterpreting utterly the natural forces amongst which they had arrived), love/hate relations (a longing to live outside the city followed by shock and indignation when confronted by a coyote) and a refusal to take seriously (or rather a belief that money – 'public' money – could and should be used to combat) a whole slew of nonhuman dynamics (from tectonic plates to river basins to bush fires). This has been a human–nonhuman negotiation of place conducted, on the human side, within an overweaning presumption of the ability to conquer. It is a manifestly different negotiation from that which has, for much of the past few hundred years, characterised an Amazonia where although in fact the interpenetration of human and nonhuman is everywhere to be found (Raffles, 2002), that interpenetration has occurred largely within an imagination of *'nature's'* overweaning power. These are extreme examples; the point is only that in every place there will be such negotiation and that these negotiations will vary. Moreover, just as in the case of the apparently more purely human negotiations, the consequences are not confined to those places alone. The nonhuman connectivities of both Los Angeles and Amazonia are global in their reach.

It is useful indeed to recognise the wider relevance of the doubts about space which first occur, to some, on the streets of the city. By that means, the import of the city is both increased and reduced. Increased, because it is, or has been, this particular kind of space which has so frequently refused to be contained within pregiven frameworks of thought and which has thus become the *espace provocateur* for more general new thinking. Reduced, because after all the city is not so absolutely special. Other doubts can be raised (and are so for me) in other places. This is important for political reasons. While the focus on cities has been productive it can be repetitive, with its insistent excited mantras, and it is excluding – not only of other, non-urban, places but of wider spatialities of global difference. It has its dubious ironies too: while globalisation is so often read as a discourse of closure and inevitability, too many of the new tales of the city are all about openness, chance and getting lost. Neither alone is an adequate story; together they are especially politically inadequate, their coexistence allowing us to play to our hearts' content on the urban streets, all the while inexorably caught up in the compound of global necessity. As King (2000) has pointedly suggested, Western academics' focus on Western world cities, the realms in which they tend to live, may be another form of inwardlookingness. Clark's argument revolves in part around material relations between Europe and Aotearoa New Zealand. In the late nineteenth century the biotic impact of colonialism was running riot: 'while the cities of the centre may have presented vistas pulsing with "the ephemeral, the fugitive, the contingent", the settler formation could offer entire landmasses convulsing with the shock of the new' (Clark, 2002, pp. 117–18). Perhaps other things could be learned by reflecting on other places.

Los Angeles and Amazonia, as they were to become, were new to the early European settlers. But even for those who do not roam so far, or even those who remain 'in place', place is always different. Each is unique, and constantly productive of the new. The negotiation will always be an invention; there will be need for judgement, learning, improvisation; there will be no simply portable rules. Rather it is the unique, the emergence of the conflictual new, which throws up the necessity for the political.

14

there are no rules of space and place

To return for a moment to the poster described in the previous chapter, depicting the immigrant boulder found in the Elbe. When the poster was put up, on a range of measures Hamburg was one of the richest cities in Europe – a wealthy city in a wealthy and powerful country. The campaign to recognise its essential hybridity, even down to the rocks, and the attempt to use this to question the terms of debate (what is local? not local?), to remove a ground from those who would argue, now, for closure (there is no appealing to an authenticity of the soil), is one which the political left is in general likely to applaud. Openness is good. 'The left', broadly speaking, deplores the closures of Fortress Europe and *la migra*. Quite right. Yet it is important to be clear about the terms of debate which underlie that position.

For at least parts of the left will also on other occasions argue equally vociferously *against* openness. While much of the language of enlightened cultural studies and the wider rhetorics of hybridity and unboundedness chime (sometimes all too easily) with the dominant tropes of neoliberalism, many of the same constituency are equally opposed to unbridled free trade: they stand against the enforced levering open of the economies of the South to Northern goods and services, opposing GATS and MAI; they defend the claim of indigenous peoples to their land and their close relation to it (all the while deploring the claim by Serbians). Some would counterpose to the triumphalism of globalisation a romanticism of the local. Just as the bulk of the political right is 'inconsistent' in extolling the free movement of capital while working actively to prevent the free movement of labour, and just as this is achieved by hailing in legitimation two contradictory geographical imaginations, so the left can often be found in the mirror, opposing both positions (arguing against free trade and for unrestricted migration) and on grounds of equally antinomic principles.

How, for instance, and in the context of the Hamburg case and the wider argument for relaxing restrictions on immigration into the European Union,

---

Marking out the heart of the Amazon

Greenpeace has just completed a month-long expedition to the Deni Indian lands in the western Brazilian Amazon. Greenpeace is working with the Deni to help get recognition of their traditional territories through the legal process of demarcation.

Deni land is under threat from WTK, a Malaysian logging giant with a string of convictions for trading in illegal logs. WTK bought over 313,000 hectares of pristine rainforest in this region of Amazonas. About half of this overlaps Deni territories, and it was sold without the Deni's knowledge or consent. In 1999, Greenpeace first made the ten-day journey from Manaus to the Deni land by riverboat to check the status of this territory.

The Deni lands are very remote and crucial to the survival of the remaining 800 Deni Indians. The Deni want demarcation in order to help keep their way of life. They live without electricity, telephones, postal service or a written language. In Brazil, once Indian land is legally demarcated it is held in perpetuity for these communities and no industrial activities are allowed in the area. Until this process is finished, the forest remains at risk.

The government process is painfully slow. The federal government sends in officials to determine the range of the community's lands, write reports and draw a map. They then contract a company to cut a six-metre border through the jungle. The Deni themselves would be side-lined in the process and it can take years.

Therefore, with the support of Greenpeace and two indigenous peoples' organisations, the Deni are pursuing the unusual step of self-demarcation. We are helping them gain information and practical skills such as the use of a GPS (satellite location device) and other technical equipment, so they can define their own territorial boundaries and take direct control of the process to force the government to act in the interests of its people and forest.Visit www.greenpeace.org.uk/amazon.htm

Courtesy of Greenpeace (http://www.greenpeace.org)

should we react to Greenpeace's campaign with the Deni of Amazonia? There are, of course, particular issues here. One of these concerns the lack of democracy in what has happened so far (see box above). We should, perhaps, be supportive of Deni participation in the future of these lands. Yet how does that square with our political response when a tabloid-saturated English populace clamours for an end to foreign immigration? Is majority local opinion always in itself 'right', or not? Or again, one might point to the fact that the rejection of the invasions of their land is necessary for the Deni 'to help keep their way of life'. But that is just what has been argued against immigration into the UK, or by middle-class villages 'under threat' from the policy of dispersal of refugees. What is certain is that there are no general spatial principles here, for they can always be countered by political arguments from contrasting cases. The 'locals' (even if they can, even provisionally, be defined) are not always 'right', nor is abiding by their majority opinion always the most democratic course to adopt.

'Defence of a local way of life' can likewise cut both ways. The question cannot be whether demarcation (boundary building) is simply good or bad. Perhaps Hamburg should indeed open up, while the Deni are allowed their protective borderlands.

Holding such apparently contradictory positions may be perfectly legitimate. It all depends on the terms on which the argument is based. When those on the right of the political spectrum argue, say, *for* the free movement of capital and *against* the free movement of labour it does not necessarily entail a contradiction. It only lays itself open to that charge (and thus open to that kind of political challenge) when each argument is legitimated by an appeal to a geographical imagination hailed as a universal, and when (as in this case) the two legitimating imaginations contradict each other. The 'inevitability' of a modern world without borders *versus* the 'naturalness' of a world in which (some) local people have a right to defend, with borders, their own local place. It is perfectly coherent to argue both for a significant relaxation of European rules on immigration (greater openness) and for the right of developing countries to put up protective barriers around, say, a vital sector of production or a nascent industry (greater closure) (see Massey, 2000a). The issue is not bounded or unbounded in itself; not a simple opposition between spatial openness and spatial closure. Not spatial fetishism.

Laclau and Mouffe, in their development of an approach to radical democratic politics, argue that 'there is no universal politics of topographic categories' (2001, p. 180). In their exemplification of this they work through debates around the party form and around the question of the state. They point out that while 'the state' in some circumstances incarnates every form of domination, in others it is an important means for effecting social and political advance. Likewise 'civil society', so often simply opposed to the state, may be at the same time 'the seat of numerous relations of oppression, and, in consequence, of antagonisms and democratic struggles' (p. 179). In other words, we cannot assume a priori that the state is 'good', civil society 'bad', or *vice versa*. Thus 'there is not *one* politics of the Left whose *contents* can be determined in isolation from all contextual reference. … all attempts to proceed to such determination *a priori* have necessarily been unilateral and arbitrary, with no validity in a great number of circumstances. … we shall never find one which does not present exceptions' (p. 179, emphasis in the original). What geographers have long criticised as spatial fetishism is in this political sphere subject to exactly the same difficulties. (And indeed Laclau and Mouffe give a rare but welcome, if rather abstract, hint of recognition of the fact that the impossibility of such a universal topography is itself a product of geography, when they write: 'The exploding of the uniqueness of meaning of the political – *which is linked to the phenomena of combined and uneven development* – dissolves every possibility of fixing the signified in terms of a division between left and right' (p. 179; my emphasis).) Abstract spatial form, as simply a topographic category, in this

for space • a relational politics of the spatial

instance openness/closure, cannot be mobilised as a universal topography distinguishing political right/left.

The argument about openness/closure, in other words, should not be posed in terms of abstract spatial forms but in terms of the social relations through which the spaces, and that openness and closure, are constructed; the ever-mobile power-geometries of space-time. Hamburg and the Deni are set within very different power-geometries, very different geographies of power. The issue is one of power and politics as refracted through and often actively manipulating space and place, not one of general 'rules' of space and place. For there are no such rules, in the sense of a universal politics of abstract spatial forms; of topographic categories. Rather, there are spatialised social practices and relations, and social power. And it is in political positions which address directly questions of that (always already spatialised) social power that answers, and they will therefore of necessity be particular answers, to (particular) questions of space and place must be sought. It is a genuinely political position-taking not the application of a formula about space and place.

Hard up against and intimately entangled with the clashing trajectories of capital in London are other conflicts. These have their roots in that other element of globalisation which derives from migratory movements and ethnic mixing. Downstream from the heartland of the financial City, the East End of London, and especially its Isle of Dogs and the surrounding boroughs, had been caught up in the maelstrom which was to produce London the twenty-first-century world city. The docks on which the area had for a century been focused were now dead. Unemployment was high, poverty endemic, vast areas of riverside land lay wasted and despoiled. The property sector had eyed the area and, through the London Docklands Development Corporation (LDDC) and with huge amounts of public subsidy, led a redevelopment which recreated the area, in part, as an extension of the City for world city industries. The story is well known, the dramas of Canary Wharf well documented.

It was not an uncontested process. In particular, during the period of the left-wing Greater London Council (1981–86), groups of working-class residents drew up, with help and encouragement from the Council, an alternative set of proposals, including a People's Plan for Docklands. One of the issues which the campaigns tried to confront was precisely that conflict between the financial world city and the other Londons, which was sketched in the previous chapter. There was a plea for 'decent working-class jobs', for sectors of production which both because of the changing nature of the economy overall and, most especially, because of the inexorable pressure on this particular part of the metropolitan land market, were going to have great difficulty in surviving without a

166

dramatic change in political commitment and policy direction. Another of the issues about which local people were concerned was incoming residents. One of the LDDC's aims was to create 'a more balanced community' (Holtam and Mayo, 1998, p. 2) (as ever, it is only working-class residential areas which appear to require dilution). The emphasis had therefore been on building private sector housing for sale, and for sale at prices well beyond the reach of people already, or recently, resident in the area. After the offering of considerable inducements (as ever, these daring risk-takers of modern capitalism don't actually like to take risks), the place slowly took on a certain cachet. What followed was portrayed as, and contested as, an invasion of yuppies. One of the terms of the contestation was that 'this is a working-class area', and the political left beyond the area, in large measure, was supportive of the cry.[9]

But there was another battle over the nature of the openness/closure of this place. Again the area was caught up in 'globalisation', but this time of a different sort. When one particular new housing project was let by the Council, using the criterion of greatest need, 28 per cent of the new properties went to people of Bangladeshi origin and white working-class people protested that 'it felt like an invasion' (Holtam and Mayo, 1998, p. 3). A resentment, with undoubtedly racist overtones, began to spread.[10] The left, in general, taking an anti-racist position, deplored the rhetorics which would attempt to enforce closure on the area.

The central stake in these two struggles took the same spatial form: 'invasion', in each case as a result of the changing imbrication of this place within capitalist globalisation, and an attempt at protective closure. What had changed from the first to the second, and what had changed the whole political nature of the issue, and the attitude of the broad left towards it, was the addition of a single word: the adjective 'white'. But if closure could not be justified in the second case by a simple appeal to the supposed (white working-class) authenticity of place, neither can it legitimately be wielded in the first by an appeal to (working-class) authenticity of place. Spatial rules (topographic categories such as openness, closure, claims for an authenticity of place) are inadequate grounds for either struggle. Once again, there can be no such a priori politics. The decision on whether or not one argues for openness, or for closure, must be an *outcome*, the result of an assessment of the specific power-relations and politics – the specific power-geometries – of the particular situation. In Docklands, the contrast in the geographies of power which lay behind the two invasions was what was crucial. The resort to general spatial principles depoliticised that contrast.

This, then, is a further aspect of our responsibility towards place; and again there are no spatial rules. Yet there is, I would contend, another issue here, which concerns the grotesque inequity of those responsibilities. When the local Council introduced a Sons and Daughters housing policy, which attempted to allow for a degree of continuity between generations in the area, it too was roundly criticised. In its wariness of the potentially racist effects of this policy, and of an exclusivist localism (but then what of the Deni?), this was in general terms a criticism which was important. Yet these are not general terms. This is an area subject to the most enormous pressure. Already an Urban Priority Area (a designation denoting desperation), with 75 per cent of households on an income of less than £7000 per annum, over half of all school children qualifying for free school meals and some of them, because of a shortage of local school places, having to be bussed elsewhere, it lay right up against the blatant display of gross wealth both in the City up the road and now here on the Isle of Dogs itself. As to housing, at the same time as the new private sector residences were going up:

> council house sales and the council's inability to reinvest in new building, had caused a withering of the council stock. 35% of white households and 47% of ethnic minority households on the Isle of Dogs were, on the Council's admission, overcrowded.
>
> In its housing allocations policy the Council's borough-wide priority has to be for those most in need, the homeless. According to the 1991 Census, 28% of the population of Tower Hamlets was Bangladeshi. On the Isle of Dogs it was 14%. A borough-wide housing lettings policy giving priority to homeless families meant that the Isle of Dogs experienced an increase in the proportion of Bangladeshis being housed there. (Holtam and Mayo, 1998, p. 2)

Holtam and Mayo, writing for the Jubilee Group of socialist Christians working in the area, go on to say, 'The Isle of Dogs in 1993 was a community which had not been listened to, and had been neglected' (p. 3) (for background to the group see Leech, 2001). To talk of 'community' begs many questions, and by this point the area was already ethnically disparate and varied in its reactions. But the feeling of neglect, and of 'not being listened to', was undoubtedly real. In September 1993, in a local by-election in the Millwall Ward on the Isle of Dogs, a member of the overtly racist British National Party was elected.

The refraction here, of class and ethnicity, of power and politics and issues of identity, through space and place, and the complex mobilisation *of* space and place as weapons as well as stakes in this knot of conflicts, is peculiarly fraught.[11] Such intensity does not confront me in (working-class, ethnically mixed) Kilburn, nor does it confront those commentators who do not live in council houses, who do not have to give back their childhood homes (quite rightly, though – as I know – it is painful) to the Council when parents die, still less does it confront the leafy suburbs (so often positively priding themselves on 'exclusivity', not needing to mobilise, explicitly, their racism and yet in wider discourses of nationhood and culture in fact underpinning it …). The clash of trajectories in this bit of east London, the spatial juxtaposition of some of the acutest antagonisms of world-citydom, is peculiarly sharp. As they tried to organise a response, the church groups found that 'all the authorities expressed the concern that they could not be seen to reward a community which had voted for the BNP' (Holtam and Mayo, 1998, p. 6). Would this area in consequence continue not being listened to?

'Cities' may indeed pose the general 'question of our living together' in a manner more intense than many other kinds of places. However, the very fact that cities (like all places) are home to the weavings together, mutual indifferences and outright antagonisms of such a myriad of trajectories, and that this itself has a spatial form which will further mould those differentiations and relations, means that, *within* cities, the nature of that question – of our living together – will be very differentially articulated. The challenge of the negotiation of place is shockingly unequal. And the politics, economics and cultures of space – through white flight, through gated communities, through the class-polarising geographies of market relations – are actively used in the production of that inequality. In the restructuring and reterritorialisation of planetary power-geometries which is the current form of globalisation, the Isle of Dogs is caught in a peculiarly complex and violent entanglement. Is this Hamburg or the Deni of Amazonia? It is neither. We come to each place with the necessity, the responsibility, to examine anew and to invent.

You arrive in Paris. Flop exhausted into a café. The distinctive mixture of coffee and black tobacco envelopes you. You anticipate some real French food. Your senses attune to the specificity of this place. Yes, this is the real Paris, France. Except of course, and you know this perfectly well at the same time, neither the coffee nor all of the food on your plate is grown in France. They're not exactly indigenous. Quintessential France is already a hybrid (just as is Hamburg, etc. etc. … as is any place). The intellectual in you knows all this; and anyway the open relational construction of place in no way works against specificity and uniqueness, it just understands its derivation in a different way.

Yet there is right now a popular movement against the invasion of this country, France, by hormone-fed beef from the USA. If 'France' (and its food) is already (always already) hybrid, does that not mean that this latest potential entrant should be allowed in too?

In August 1999 José Bové, along with a crowd of some three hundred, systematically dismantled a branch of McDonald's which was being constructed in Millau in the *département* of Avéyron. The action and the subsequent trial and sentencing became the focus of a *cause célèbre*. For Bové, and for his co-leader François Dufour (General National Secretary of the French Farmers' Confederation), the choice of McDonald's was symbolic of 'economic imperialism': 'the dismantling had been a symbolic protest against multinationals like McDonald's taking over the world' (Bové and Dufour, 2001, pp. 13 and 24).

169

One of their earliest, and probably continuing, difficulties was to distance themselves from a groundswell of support which played through easier emotions and which leapt to interpret their actions in terms of anti-Americanism in particular and nationalistic closure more generally. (Against yet another *défi américain*.) Bové and Dufour have gone to great lengths to refute these interpretations (and maybe, even, that need to deny them has helped propel their own position which has certainly become more complex and sophisticated over the years).

On the first charge, their actions themselves have been insistent. At the very moment of Millau, Dufour was planning an intervention at an American film festival at Deauville where he

> wanted to explain to the American Festival-goers that it was not their culture we objected to: that it was very welcome in our regions, but that the multinational companies had to respect our differences, our identity. We don't want hormones in our food; they're a risk to public health, and go against our farming ethics. At a more fundamental level, imposing hormones on us means that our freedom of choice in the food and culture we want is seriously restricted. Agricultural exchanges have existed for a long time: we don't advocate exempting agriculture from the politics of international trading, but we want something different from freedom of the market and the liberal economy. (Bové and Dufour, 2001, pp. 20–1)[12]

They have, moreover, made many links with like-minded farmers' groups in the USA.

The immediate spark that provoked Millau was the US surcharge of 100 per cent on imports of Roquefort cheese. The European Union's refusal to import US hormone-fed beef had been declared by the WTO to be against its rules and a time limit had been set for its lifting. When the EU failed to comply the USA retaliated with a series of surcharges of its own. Among them was one on Roquefort, and in south Avéyron 'solidarity on the issue of ewes' milk is taken for granted' (2001, p. 3). This was moreover a region with a history of organised militancy and a strong presence of 'alternative' farming stemming from the battle to prevent military expansion on the Larzac plateau over twenty years before. By the time of Millau, and even more so subsequently, the campaign embraced a nexus of issues circling around the character of the negotiation with the nonhuman through farming (against intensive monoculture and control by multinational corporations), questions of health and of the quality and variety of food, and the preservation of diversity. Farming itself is understood in an explicitly relational way: between human and nonhuman and as articulating economic, social and environmental practices and concerns. It is emphatically not just an economic activity.[13]

This is not a politics which is arguing for national closure as any kind of general principle. Bové and Dufour are also insistent that they are not opposed to globalisation in its general sense. In spite of what have clearly been difficulties stemming from their situation as farmers within the European Union they have struggled to define a position which leaps across those boundaries and builds an internationalism through alliances with other groups of small farmers the world over (such, for instance, as are brought together under the umbrella of *Via campesina*). They talk of a 'farmers' Internationale'. Their opposition is to the character of the current form of globalisation, with particular antagonisms constructed around the nature of the flows which it embodies, and the complex of relations in which they are embedded and which give them such overweening power, and – most especially – the lack of democracy in their construction. The call at this level is, among other things, for democratic control of the WTO. Clearly, then, this is not a politics of closure. What is at issue is the *nature of the relations of interconnection – the map of power of openness*. French food can continue its long history of absorbing new influences: the question is which ones, why and on whose terms.[14]

And yet … this campaign is also pro-local. It does call for a specific geography – one which values local specificity. The long quotation above gives a hint of this. But *how* is one to be pro-local? On what terms? In the actions, speeches and writings of Bové, Dufour and the other protagonists in this campaign, you can feel them struggling, often insightfully and creatively, with the terms on which *in this particular set of issues*, 'the local' can be defended. In general, they are careful not to resort to a simple nostalgia for an edenic past; what they are about is the 'farm of the future'. They recognise that localities are 'made', but are sensitive to the longevity of social structures in many rural areas (they write of 'the ties that bind' – p. 56; and the fact that 'people don't want to be uprooted' – p. 27). The local specificity which they evoke is one derived in part from the variations within 'nature'. And part of their argument is that, for them, a politically acceptable negotiation with nature would involve responding to the local variations in its rhythms (they speak frequently of rhythms): 'In intensive farming the object is to adapt the soil to the crop, never the other way round' (p. 67). Their aim, precisely, is to do it the other way around. This is a respect for local specificity, and an argument in favour of its recognition, that, in general, avoids romanticism. It recognises the place-specific conjunctions of human and nonhuman trajectories and its politics addresses the terms of their intersection. There is also a complementary theme in their argument which favours geographical diversity in itself (that diversity, variability, choice, are themselves positive goods).

And yet somehow there are still difficulties. Perhaps some of these can be gleaned from the following section in which Bové and Dufour in turn address the thorny issue of what, exactly, is meant by '*malbouffe*', and why they are against it. (In English the term is most commonly, though inadequately, rendered as 'junk food'.)

> *Bové:* 'Malbouffe' implies eating any old thing, prepared in any old way. …
> For me, the term means both the standardization of food like McDonald's –
> the same taste from one end of the world to the other – and the choice of food
> associated with the use of hormones and GMOs, as well as the residues of
> pesticides and other things that can endanger health. So there's a cultural
> and a health aspect. Junk food also involves industrialized agriculture – that
> is to say, mass-produced food; not necessarily in the form of products sold
> by McDonald's, but mass-produced in the sense of industrialized pig-rearing,
> battery chickens, and the like. The concept of 'malbouffe' is challenging all
> agricultural and food-production processes. …
>
> *Dufour:* Today the word has been adopted to condemn those forms of agri-
> culture whose development has been at the expense of taste, health, and the
> cultural and geographical identity of food. Junk food is the result of inten-
> sive exploitation of the land to maximize yield and profit. (pp. 53–4)

This is a definition which beautifully captures the relations within which
malbouffe is caught up, and which Bové and Dufour oppose. But what is the
'geographical identity of food'? In an age when even the UK Foreign Secretary
feels able to observe that chicken tikka marsala is a British national dish, this is
a difficult concept to conjure with.[15] Elsewhere, there is talk of the defence of
'the practice of an agriculture linked to one product and one area' (p. 77) (single-
product monoculture? – the local roots in Roquefort country are surely evident
here!) and claims that 'The people who live in an area have to decide how its
resources are to be used' (p. 134).[16] This last commitment does not recognise the
democratic claims which derive from wider connectivity; and much talk of
'local solidarity' also skirts the potential for conflicts within place.

My point here is absolutely not to perform some intellectual critique. Quite
the contrary. Rather it is to stress just how genuinely difficult it is *not* to resort
to an a priori politics of topographies. It is far more complicated to carry such
an injunction into practice in the formation of a particular politics than it is to
write about it as a general proposition. But, as the development of the arguments
of the *Confédération paysanne* themselves exemplify, the very effort *not* to resort
for legitimation to such topographies (local is good because it's local) is also
immensely politically productive. It forces one into the excavation of what are the
real political issues in this (particular) situation. And this in the end will resolve
itself around political antagonisms: concerning a commitment to democracy –
economic as well as political, and therefore for/against the current practices of
multinational capital – or the ethics of a particular relationship to nature, or the
significance of maintaining diversity.

There is a particular thread which runs through this bundle of debates. It comes
perhaps especially from feminists, and it cautions against an over-excited

celebration of openness, movement and flight (in the sense of escape). Catherine Nash (2002) has written about the potential validity, in political terms, of some of the pulls towards settledness and even closure in the context of the social construction of the identity of place and of the rich ambiguities of 'genealogical identities'. Susan Hansen and Geraldine Pratt warn against a new orthodoxy of exile, marginality and openness which might serve only to reinforce in new guises individualism and elitism (Pratt and Hansen, 1994; see also Pratt, 1999). Caren Kaplan (1996) has analysed the conditions which lie behind (some) postmodernisms' evocation of nomadism, the persistent attraction to 'the desert', and so forth. She points to the roots of these characteristics in aspects of the modernism which they were precisely trying to escape: how so much of this postmodern/post-structuralist literature advocates a strategy of escape which harks back to the modernist romance of the writer in exile, how that in turn plays into an (implicit) understanding of detachment as a precondition of creativity, and of distancing as a requirement for the production of knowledge. (The spatiality of knowledge production again.) She points, too, to the contrast between the individualised line of flight and the historical setting of mass migration, its conditions and the attempts to rein it in. The figures of the desert and the nomad, she argues, are – along with the other sites to which we might flee – precisely the places of the modernist Western other. They are landscapes figured through imperialist myth (and, one might add, striated *into* 'desert', 'sea' and so forth, through particular practices). They function in these discourses only through (and precisely as a result of) the Euro-American modernist imagination: 'Constructing binaries between major and minor, between developed and undeveloped, or center and periphery, in Deleuze and Guattari's collaborative texts modernity provides borders and zones of alterity to tempt the subversive bourgeois/intellectual' (Kaplan, 1996, p. 88). In that guise, such other people and places cannot have trajectories of their own; they function, Kaplan argues, 'simply as a metaphorical margin for European oppositional strategies, an imaginary space, rather than a location of theoretical production itself' (p. 88). This is, in other words and in the terms of my argument here, a failure of the imagination of coevalness. It denies a space of *multiple* becomings: the 'others' are not allowed a life of their own. As Cindi Katz puts it, it 'leaves the "minority's" subjectivity suspiciously in the lurch' (1996, p. 493; see also Jardine, 1985; Moore, 1988). And, continues Kaplan, it is also a rhetoric, and an advocacy, which fails to recognise its own (relatively powerful) subject position, for 'these spaces of alterity are not the symbols of productive estrangement or disengagement for any other subjects. These imagined spaces are invested with subversive or destabilizing power by the "visitors", as it were' (1996, p. 88). Miller has raised concerns similar to those of Kaplan but in the context of anthropology, arguing that Deleuze and Guattari's procedure commits them to an 'anthropological referentiality' which is open to criticism on both empirical and practical grounds (Miller, 1993, pp. 11–13; and see the response by Patton, 2000).

A further set of arguments revolves around the fact that both openness and closure, and both classic territory and rhizomatic flow, can be the outcome of sedimented and unequal power-relations. In Castells' evocation of a transition from a space of places to a space of flows the latter is no less 'closed' in relation to control and potential change than is the attempted closure of the nation-state. Settledness and flow, likewise, are conditions for the existence of each other. As the evolving arguments of the *Confédération paysanne* and of José Bové make clear, of equal importance as any questions of openness/closure are the mobile power-geometries of the relations of connection. Or again, the big battles of global politics in the twenty-first century look set to be equally against power-invested flows on the one hand and against closure against flows on the other. Equally, in the schema of Deleuze and Guattari, 'smooth space' is not devoid of organising power:[17] 'The multinationals fabricate a kind of deterritorialized smooth space …' (1988, p. 492); 'the smooth itself can be drawn and occupied by diabolical powers of *organization*' (p. 480; emphasis in the original). And so forth. Bruce Robbins' (1999) analysis of *The English Patient* by Michael Ondaatje confronts precisely these issues. On the one hand there is a refreshing scepticism about the nation-state and the enclosures of 'home' as loci of identity and loyalty, and a more unusual refusal to equate that home with 'woman'; on the other hand there is, as Robbins puts it, 'a tangible reminder that alternatives to domesticity do not always improve upon it' (p. 166). Simply saying 'no' to nation, home, boundaries and so forth is not in itself a political advance (it is spatial fetishism to think it will be) – in the novel the Europeans, in the name of mobility and unboundedness, casually and symptomatically invade 'a half-invented world of the desert' (Ondaatje, 1992, p. 150; see Robbins, 1999, p. 166).

Indeed the most excited embraces of flight, hybridity, openness and so forth depend upon, are motivated by, their implicit retention of a definition of closure, or authenticity, or whatever, which is anyway impossible. Thus Kaplan relates an 'exilic, melancholic romance with "distance"' to 'a strong attachment to its opposite – a metaphysics of presence' (1996, p. 73). And Donald draws out a similar argument in his reading together of Raymond Williams and Salman Rushdie: on the one hand 'Williams's excessive investment in community' and on the other 'Rushdie's possibly equally excessive celebration of migration' (1999, p. 150). 'Each', he suggests, 'is an experiential and political strategy for dealing with the (more or less conscious) loss of the *possibility* of home with which we live' (p. 150).[18] That closure of the imagined 'home' is anyway *impossible*. Deleuze and Guattari in their attachment to a bipolarity of smooth and striated can evoke a similar opposition. Thus Hardt and Negri, in *Empire* (2001), which draws on Deleuze and Guattari, on occasions exhibit this characteristic. In their advocacy of a rhizomatic politics the conceptual backcloth of smooth space has problematical effects in two ways. First, in an uneasy slippage between individual and multitude, with nothing much offered by way of addressing the negotiation of political identities in practice; no serious way of getting to grips with the heterogeneity *within* the multitude – and smooth space *is* heterogeneous. So in this political sphere one of the crucial issues is how political constituencies are formed, and how they interrelate, within this. But – and second – this smooth space also relies on its opposite, and this is equally politically debilitating. Thus Hardt and Negri fall into the trap which Kaplan and Donald detect (and which elsewhere they try to evade – see 2001, pp. 43–6); they write that 'Doreen Massey argues explicitly for a politics of place in which place is conceived not as bounded but as open and porous to flows beyond, …. We would contend, however, that a notion of place that has no boundaries empties the concept completely of its content' (2001, p. 426). We are left, there-fore, again, with two romances which are simply opposed to each other. Both the romance of bounded place and the romance of free flow hinder serious address to the necessary negotiations of real politics.

Barnett (1999), drawing on a more Derridean formulation, puts the point well: 'One lesson of deconstruction is that the political value of either fixed meaning (of closure or of identity) or of maintaining instability (of ambivalence or of difference) is not open to prior, conceptual determination' (p. 285). Indeed, as he also points out, relations of dominance may be maintained precisely through the instabilities of meaning. Feminists have often pointed to the chains of loosely linked and occasionally contradictory binaries through which oppres-sive discourses can be reproduced. The very slipperiness is one of the resources which produce the effects of power. The shifting between contradictory geographical imaginations, all of them less stable than they are claimed to be, can be an equally significant manoeuvre (see Chapter 8). The closed geo-graphical imagination of openness, just as much as that of closure, is itself irretrievably unstable. The real political necessities are an insistence on the recog-nition of their specificity and an address to the particularity of the questions they pose.

We are always, inevitably, making spaces and places. The temporary cohe-sions of articulations of relations, the provisional and partial enclosures, the repeated practices which chisel their way into being established flows, these spatial forms mirror the necessary fixings of communication and identity. They raise the question of a politics towards them. In his essay *On cosmopolitanism and forgiveness* Derrida (2001) addresses the concept of hospitality, a concept which, he argues, evokes 'not simply one ethic amongst others' but the whole question of our throwntogetherness: 'it is a manner of being there, the manner in which we relate to ourselves and to others, to others as our own or as foreigners, *ethics is hospitality*' (pp. 16–17, emphasis in the original). The occasion is the International Parliament of Writers in Strasbourg in 1996 and the politi-cal focus is asylum seekers and refugees (the Parliament was proposing that there be cities of refuge – *villes franches, villes refuges*). The logic of the argument, however, is that of openness/closure more generally. On the one hand there must be recognition of an unconditional law of hospitality: unrestricted openness.

On the other hand there is the differentiated reality of the need for conditionality. As Simon Critchley and Richard Kearney put it in their Preface: 'these two orders of the unconditional and the conditional are … in a relation of contradiction, where they remain both irreducible to one another and indissociable' (Derrida, 2001, p. xi). 'All the *political* difficulty of immigration consists in negotiating between these two imperatives' (p. x, emphasis in the original): the 'moment of universality that exceeds the pragmatic demands of the specific context' but where such unconditionality is not allowed 'to programme political action, where decisions would be algorithmically deduced from incontestable ethical precepts' (p. xii). In Derrida's own words, we have to operate:

> within an historical space which takes place *between* the Law of an unconditional hospitality, offered *a priori* to every other, to all newcomers, *whoever they may be*, and *the* conditional laws of a right to hospitality, without which *The* unconditional Law of hospitality would be in danger of remaining a pious and irresponsible desire, without form and without potency, and of even being perverted at any moment.
>
> *Experience and experimentation thus* (pp. 22–23; emphases in the original).

15
making and contesting time-spaces

A number of years ago I embarked on a research project which engaged with two contrasting kinds of time-space: the scientific laboratory and the home.[19] The high-tech scientists who worked in the laboratories were in private sector R&D; they were whizz kids of modern economic development, with high status and high rewards, and 95 per cent of them in the UK as a whole at that time were male. The laboratories were in stylish modern buildings on a science park or, more rarely, in a converted, still stylish, older building. The dominant imaginative geographies of such places are tied up with globalisation and with the 'new economy': these are among the most globalised parts of the economy, and the spaces they inhabit are imagined as equally open and flexible, set in a mobile global information system advertised as being in the vanguard of breaking down old rigidities. And certainly, as we began to explore these places, they seemed to live up to the image. Every day the activities here were hooked up with activities on other continents: conference calls, emails, intellectual exchange and contract negotiations. Trips abroad were routine. Truly globalised places, nodes of international connectivity even more than local (and mirroring in the nature of their own globalisation, indeed producing it in part, the structural inequality within the wider phenomenon). In these senses, then, these high-tech workplaces were the epitome of openness. Moreover, at night, usually quite late and after a long day, our research scientists left their globalised laboratories to go home. And a goodly number of them went home to a country village (we were focusing on the Cambridge area), to a converted cottage with a garden: the English emblematic home. It was, it seemed as we set about our research, a classic return from globalised days to a bounded local security.

Such a contrast would have important resonances. First (and this point will not be undermined by the surprises which the research threw up) it instantiates at a local level and at the level of individual lives that emerging characteristic of globalisation as we know it whereby 'the powerful' (through whatever source their power derives) have the ability both to conduct and control their lives internationally and to defend a secure place of their own. And, second, it resonates with that other story, of male mobility and female enclosure, of which so many have written. There seemed to be a clear cartography of gender and a classic contrast between global openness and local self-containment.

The beauty of empirical work is that you have no sooner reached such neat and satisfying conclusions than they start to exhibit cracks and queries. The more we were in those laboratories the more their *closure* impressed itself upon us. Their devotion to a highly specialised activity (thinking; 'research and development'), their very design as celebrations of that activity. Where other kinds of practice were present (the kitchen, the table-tennis table) they were there in order to increase the effectiveness of this time-space in enabling the performance *of* this single-minded activity. There was something strange, sometimes, about being in these time-spaces. They were quite spare and sparse, with little evidence of the rest of life; no supermarket bags spilling groceries, no non-work reading matter. Single-minded spaces. None of the places we visited had a crèche; in one of them workers' children were kept out, even at week-ends, by security guards (a child had once, it seems, behaved inappropriately). And security guards defended some of the laboratories more generally. Glob-alised places, indeed, but selectively so; open only to a highly particular set of practices and to similar others. They, and the science parks on which they are so often set, are (as was seen in Part *Four*) the product of the intertwining of trajectories with great historical and geographical reach, and those trajectories themselves are part of the production of, and the conditions for, the terms of current closure. These globalised workplaces are specialist and excluding spaces, defensive, quite tightly sealed against 'non-conforming' invasions from other worlds. Such closures are constructed both materially and imaginatively, through both security guards and the symbolisms of exclusivity. Their very existence as specialised places of R&D (geographically removed from physical production) both is a product of and simultaneously reinforces the idea of the necessity for a space of Reason, defended against contaminations of the Body. The clipped modernity or the rural chic, the landscaping which reflects long histories of the generation of 'taste' and class distinction, contribute to the status and success of these places: the negotiation with the nonhuman is geared to reinforce the exclusivity. It is, of course, a closure which as ever and even in terms of its own restricted dimensions, is impossible to hold (see Massey, 1995b; Seidler, 1994) but it is effective enough in moulding the identity of the ('logical', 'masculine') scientist, in reinforcing the cachet of their profes-sion, and in underpinning the legitimacy and status of a particular kind of knowledge.

Such thoughts made us look in a different way too, as we carried on with our interviews, at the homes of these scientist-researchers. It was not that the terms of the contrast between the two time-spaces (openness/closure) had simply reversed; but the nature of the contrast had certainly shifted. The homes now seemed in some ways relatively open and porous spaces. Clearly entry was carefully restricted, guarded against a whole range of unwanted potential intrusions. Yet in comparison with the tunnel-vision specialisation of the labs these houses were a base for a variety of people, for multiple interests and activities, and they were littered with evidence of this multiplicity and variety. Specifically, too, while the laboratories were definitively not invaded by domesticity, these homes were certainly invaded by 'his' work. There were scientific journals on the settee, by his chair. There were the myriad virtual invasions, recounted in detail and at length, by both scientists and their (female) partners, of his thinking about work while playing with the children or, on a day out, tales of keeping notebooks by the bed in case of a good idea, of worrying about work in the bath. Often, too, these variegated time-spaces which were homes had studies within them, where he would retire to work. And these places-within-places would be constructed much more along the lines of the lab. This was Daddy's office, you didn't go in there; an inner sanctum (see also Wigley, 1992). There was a decidedly one-way invasion (one which rather casts in a different light the usual rhetoric of some unspecified blurring of the boundaries of home and work); an invasion of home by work but not *vice versa*, and the research went on to investigate why the one time-space was so much 'stronger' than the other.[20]

The point here, though, is rather to ponder the nature of all this openness and closure. Each of these time-spaces is relational. Each is constructed out of the articulation of trajectories. But in each case too the range of trajectories which is allowed in is carefully controlled. And each time-space, too, is continually shifting in its construction, being renegotiated. In middle-class Western homes like these there is an ever-increasing presence of commodities drawn from around the world and a huge variety of interconnectedness through new communications technologies; but there is also talk of a retreat to the privatised, individualised, nuclear family and a regrowth of gated communities. Some borders are being dismantled, some renegotiated, and yet others – new ones – are being erected. The real socio-political question concerns less, perhaps, the *degree* of openness/closure (and the consequent question of how on earth one might even begin to measure it), than the *terms* on which that openness/closure is established. Against what are boundaries erected? What are the relations within which the attempt to deny (and admit) entry is carried out? What are the power-geometries here; and do they demand a political response?

Aldo van Eyck's 'fundamental belief' is said to have been that 'a house must be like a small city if it's to be a real home; a city like a large house if it's to be a real home' (Glancey and Brandolini, 1999). That is an amazingly challenging proposition. On the one hand how could a home be like a city when, as we so constantly aver, cities are precisely arenas of chance encounters. (And yet that thought itself should also bring to mind the countless exclusions which together accumulate to produce that space of the city.) On the other hand, that *is* one of the characteristics of space; that it is the condition of both the existence of difference and the meeting-up of the different. (Yet that is so often too much for us: the challenge of space can rarely be met full on.) The current form of social organisation of the time-spaces both of the scientific laboratory and of the home are precisely attempts to regulate, though in very different ways, the range and nature of the adventures and chance encounters which are permissible. Each is a way of dealing with the multiple becomings of space. Developing a relational politics around this aspect of these time-spaces would mean addressing the nature of their embeddedness in all those distinct, though interlocking, geometries of power. If entities/identities are relational then it is in the relations of their construction that the politics needs to be engaged. In the case of the laboratories, the politics might lie in addressing how these 'privileged scientific sites' (Smith and Agar, 1998) are produced through and productive of an understanding of certain forms of knowledge as legitimate, in addressing the constitution of certain forms of masculinity; and in addressing how these are cross-cut by the spatialisations of capitalist competition and their repercussions back on the process of production of knowledge. In other words, it would involve a politics towards those trajectories pointed to in Part *Four*. The closures of the nuclear family home can be opened to a critique parallel to that now so commonly made of those other old conservative enclosures, the nation-state and the local community. And so forth.

And yet what van Eyck was after, at least in his early years, was to create spaces where you *might* come upon the unexpected, have chance encounters (that mixture of order and accident which, as we have seen, he called 'labyrinthine clarity'). James Donald (1999) pursues a similar idea as he thinks through what might be a way of 'doing architecture differently' for the city – an architecture which both acknowledges the past (its 'critical power of remembering in grasping urban space as historically and temporally layered' – p. 140) and is as open to an unknown, and through architecture indeterminable, future. It might be an architecture which 'attempted to *build in* flexibility, tolerance, difference, restlessness, and change' (p. 142; emphasis in the original) (Donald here is writing of Tschumi). Andrew Benjamin (1999) has made a similar point as a more general proposition, that 'architecture can avoid the traps of prescriptive formmaking whilst releasing the potentials of the incomplete, of the yet-to-be' (Till, 2001, p. 49). In fact, there *will* be adventures however the space is designed, whether it be laboratory, home, or the urban park. The chance encounter intrinsic to spatiality cannot be totally obliterated. It is (in part) this indeed that makes time-spaces, however much we try to close them, in fact open to the future; that makes them the ongoing constructions which are our continuing responsibility, the ongoing event of place which has to be addressed.

A relational politics of place, then, involves both the inevitable negotiations presented by throwntogetherness and a politics of the terms of openness and closure. But a global sense of places evokes another geography of politics too: that which looks outwards to address the wider spatialities of the relations of their construction. It raises the question of a politics of connectivity.

There is a host of issues here: it questions any politics which assumes that 'locals' take all decisions pertaining to a particular area, since the effects of such decisions would likewise exceed the geography of that area; it questions the predominance of territorially based democracy in a relational world; it challenges an all-too-easy politics which sets 'good' local ownership automatically against 'bad' external control (Amin, 2004). It raises the issue of what might be called the responsibilities of the local: what, for instance, might be the politics and responsibilities towards the wider planet of a world city such as London?

It also reinforces that argument that it is no response to globalisation simply to press the case of the local. The political meaning of 'local' cannot be determined outside of specific contextual reference. Local/global in itself cannot be an adequate surface along which to constitute political antagonism. The political questions become not *whether* globalisation but what kinds of interrelations are to construct an alternative globalisation, and thus not simply a defence of place-as-is, but the political project of the nature of the places within it. Paul Little, in exploring 'globalization and the struggles over places in the Amazon' tries precisely to steer this course: the 'most pressing questions become: what type of globalization do we want? And what kind of places should this process be creating?'. To address these questions, he stakes out three propositions: first that social justice criteria must be used for the political legitimation for these historical claims to Amazonian places (in other words, not supposedly universal spatial claims); second that Amazonia is already mixed ('Colonists, miners, fishermen, urban dwellers, and factory workers …') and that the resultant variegation of these places requires explicit political attention; and third that there needs to be a creative relation to the nonhuman as another participant in this making of places (places are not just human constructs): 'the current hegemonic notion that the biophysical environment is nothing more than an inert mass that humans can manipulate and dominate must be abandoned, and replaced with the notion that it too is an essential actor, albeit a natural and not a social one, in the creation of liveable places' (Little, 1998, p. 75).

And yet, of course, most struggles around globalisation are inevitably 'local' in some sense or other. A long tendency on the left has been either to denigrate them for being 'only local' or to romanticise them for their supposed rootedness and authenticity. There are spatial imaginaries in play here: both responses depend upon a notion of the local as effectively closed, self-constitutive. The political question of how to move beyond the single local struggle can then only be addressed through some imagination of an accumulation of localisms: the mere adding up of particularities. Each local struggle is already given, internally generated, with the consequence that their accumulation is intended to involve no change in their nature; indeed the very process of 'adding up' is often viewed warily, as a potential threat to local authenticities. Pre-existing conflicts between different local demands might on this reading hinder the achievement of each of them individually. Neither a concept of the local as 'only local' nor a romanticisation of the local as bounded authenticity, in other words, offers much hope for a wider politics.[21]

The topography is very different when the local (and, concomitantly, the global) is thought relationally. Then each local struggle is already a relational achievement, drawing from both within and beyond 'the local', and is internally multiple. As Featherstone (2001) argues, even 'militant particularisms' are openly and relationally produced. The potential, then, is for the movement beyond the local to be rather one of extension and meeting along lines of constructed equivalence with elements of the internal multiplicities of other local struggles. The building of such equivalences is itself a process, a negotiation, an engagement of political practices and imaginations in which ground is sought through which the local struggles can construct common cause against a (now differently constructed) antagonist. And this ground will itself be new; politics will change in the process. Moreover, within that process – precisely through the negotiation of a connection and the constitution of a common antagonist – the identities of the constituent local struggles are themselves subject to further change. As Laclau and Mouffe have put it, equivalence 'does not simply establish "an alliance" between given interests, but modifies the very identity of the forces engaging in that alliance' (2001, p. 184). Using a different terminology, and developing the ideas of transversal politics (Yuval-Davis, 1999), Cynthia Cockburn writes of 'alliances holding together differences whose negotiation is never complete, and is not expected to be so' and in which the negotiations themselves are productive of political, and personal, identity (Cockburn, 1998, p. 14). Such an alternative topography for thinking local/global by no means indicates a politics which is easy to prosecute but it can help to get a grasp on the – potentially politically productive – tension between equivalence and autonomy (the continuation of distinctiveness within a constructed relatedness) and it is also a topography which, in keeping with the arguments of Chapter 14, rather than providing a template of answers forces the posing of questions about each specific situation.

Such an understanding entirely reworks formulations such as 'the relation between local and global'. What is involved is an immensely difficult, always grounded, and 'local' if you like, negotiation. One effect is to demand far more of the agents of local struggle in the construction of both identity and politics than there is room for in that topography where identity seemingly emerges from the local soil. Theorists of radical democracy, on the other hand, have rarely engaged with the complexity and real difficulty of this construction of equivalences. Dave Featherstone (2001), in a whole range of studies has emphasised and explored precisely this, showing in detail how the identities of political constituencies are constantly produced through negotiation at the intersection of a nexus of connections. The experience of the *Confédération paysanne* is similar:

> We didn't expect one side to convince the other. In any case these positions aren't so different as they may seem, because they're united in their assessment of the harm done by the WTO. You can't talk about factions within *Via campesina* … What holds for Santiago or Bamako doesn't necessarily hold for Rome or Paris. The exchange of opinions and experiences makes this a wonderful network for training and debate. (Bové and Dufour, 2001, p. 158)

> The strength of this global movement is precisely that it differs from place to place, while building confidence between people. (p. 168)

> Actions can change the ideas of those who take part. (p. 170)

All this is integrally, and significantly, spatial. The differential placing of local struggles within the complex power-geometry of spatial relations is a key element in the formation of their political identities and politics. In turn, political activity reshapes both identities and spatial relations. Space, as relational and as the sphere of multiplicity, is both an essential part of the character of, and perpetually reconfigured through, political engagement. And the way in which that spatiality is imagined by the participants is also crucial. The closure of identity in a territorialised space of bounded places provides little in the way of avenues for a developing radical politics.

Yet there is a prevailing attitude towards place that works against that kind of shift of political gear. Spatial imaginaries both in hegemonic and counter-hegemonic political discourses, and in academic writing, hold it back. Of prime importance here is the persistent counterposition of space and place, and it is bound up with a parallel counterposition between global and local (although as Dirlik points out the two pairings can be distinguished). Over and again, the counterposition of local and global resonates with an equation of the local with realness, with local place as earthy and meaningful, standing in opposition to a presumed abstraction of global space. It is a political imaginary which, in a range of formulations, has a powerful counterpart in reams of academic literature. In one of the founding geographical statements of this ilk, Yi-Fu Tuan proposed that '"space" is more abstract than "place"' (Tuan, 1977, p. 6). Philosopher Edward Casey asserts that 'To live is to live locally, and to know is first of all to know the places one is in' (Casey, 1996, p. 18), and social theorists not infrequently aver that 'Place is space to which meaning has been ascribed' (Carter et al., 1993, p. xii). It is, for me, the real difficulty of Heidegger's reformulation of space *as* place (which would in principle seem to point in the right direction): in the end, Heidegger's notion of place remains too rooted, too little open to the externally relational. And terminologically, the *effect* of this focus has been to reinforce a space/place counterposition. It works against the notion of place proposed in Part *Four*.

Perhaps the most difficult context within which this issue arises is aboriginal culture – since the claim there so often made is the inseparability of life and land. A special issue of the journal *Development* (volume 41, number 2, 1998) is devoted to a thoughtful, and very varied, wrestling with this problem. Arif Dirlik, for example, calls for 'conceiving of place as a project' (1998, p. 7) and is well aware of the fact that this is a politically tricky proposition (being appropriatable across the political spectrum). There is an insistence on the phrasing 'place-based' rather than 'place-bound', which is important because it recognises the relations of space beyond place. Yet the frequent claims that 'Place consciousness … is integral to human existence' (Dirlik, 1998, p. 8) still nag. Why such essentialism? There is no *need* in these arguments to press the claim to a universal; and in many ways such a claim runs counter to the tenor of the rest of the analyses.

Finally, the counterposition is sometimes set in a wider context:

> The move from tangible solidarities understood as patterns of social life organized in affective and knowable communities to a more abstract set of conceptions that would have universal purchase involves a move from one level of abstraction – attached to place – to another level of abstraction capable of reaching out across space. … The shift from one conceptual world, from one level of abstraction, to another, can threaten the common purpose and values that ground the militant particularism achieved in particular places. (Harvey, 1996, p. 33, cited in Featherstone, 2001)

All this, to my mind, rests upon a problematical geographical imagination. To begin with, it is to confound categories. The couplets local/global and place/space do not map on to that of concrete/abstract. The global is just as concrete as is the local place. If space is really to be thought relationally then it is no more than the sum of our relations and interconnections, and the lack of them; it too is utterly 'concrete'. (It is evident here how romanticising the local can be the other side of understanding space as an abstraction.) Nor is the elision of meanings of 'universal' helpful for it manages both to romanticise the local and to instate the global (as the abstract universal) as either the only real struggle to be aimed at or as so ungrounded and 'up there' as to be unaddressable (see Massey, 1991b; Grossberg, 1996). It is bound up with, is yet another geography of, that dualism between Emotion (place/local) and Reason (space/global).

An understanding of the world in terms of relationality, a world in which the local and the global really are 'mutually constituted', renders untenable these kinds of separation. The 'lived reality of our daily lives' is utterly dispersed, unlocalised, in its sources and in its repercussions. The degree of dispersion, the stretching, may vary dramatically between social groups, but the point is that the geography will not be simply territorial. Where would you draw the line around the lived reality of your daily life? In such approaches words such as 'real', 'everyday', 'lived', 'grounded' are constantly deployed and bound together; they intend to invoke security, and implicitly – as a structural necessity of the discourse – they counterpose themselves to a wider 'space' which must be abstract, ungrounded, universal, even threatening. Once again the similarity between the conception of information as disembodied and of globalisation as some kind of other realm, always somewhere else, is potent. A technology-led understanding of globalisation reinforces the connection. It is a dangerous basis for a politics. One cannot seriously posit space as the outside of place as lived, or simply equate 'the everyday' with the local. If we really think space relationally, then it is the sum of all our connections, and in that sense utterly grounded, and those connections may go round the world. Indeed, Harvey elsewhere makes exactly this point: 'In modern mass urban society, the multiple mediated relations which constitute that society across space and time are just as important and "authentic" as unmediated face-to-face relations' (Harvey, 1993, p. 106, cited in Corbridge, 1998, p. 44). It is not necessary to sign up to distinctions between mediated and unmediated to agree with the intention here. As Hayles writes of information, 'it cannot exist apart from the embodiment that brings it into being as a material entity in the world; and embodiment is always instantiated, local, and specific' (1999, p. 49). Does the argument that place is space which has been endowed with meaning not allow those stretched relations of a globalised world to have meaning too? My argument is not that place is not concrete, grounded, real, lived etc. etc. It is that space is too.

The difficulties of making this argument politically effective are reinforced by notions of the global as 'out there' or 'up there', not needing, in the rhetorics of Gates (1995) and Negroponte (1995), to touch ground. They are reinforced by imaginations of place, or of the local, as victims of global space: the association, in Escobar's (2001) words, of place, the local and vulnerability on the one hand, and space, capital and agency on the other (Part *Three*).

And there are other issues too. It does seem so difficult to remember, in the restaurant say, the complex of far-flung relations through which the mange-touts arrive on your plate. In the now famous words of John Berger, 'It is now space rather than time that hides consequences from us' (1974, p. 40). Some of this difficulty may be the result of the still-remaining impact (in this world said to be increasingly 'virtual') of material juxtaposition; of sheer physical proximity. There are, too, all the rhetorics of territory: of nation, family, local community, through which we are daily urged to construct our maps of loyalty. (While other rhetorics simultaneously persuade us that this is the age of far-flung connectivity. It is that spatial double-think which was encountered in Chapter 8, that conflictual spatiality of the attempt to combine neoliberalism with conservatism, which has typified, and disrupted, the rhetorics of Thatcher, Blair, Bush, Clinton and many more besides.) There is, connectedly, the fact that our formal politics is organised territorially (in this world so often called a space of flows). Some of the difficulty may be intimately (the apposite word) connected to a cultural obsession with parent–child relationships, the focusing of the question of care primarily within family relations (Robinson, 1999). Why do we so often and so tightly associate care with proximity? Even those who write of care for the stranger so often figure that relationship as face-to-face. It is the counterpoint perhaps to the persistent lack of acknowledgement of the strangers who have always been within.

The constructedness of these attitudes is evidenced, as ever, by their spatial variability and their historicity. Lester (2002) has excavated, through debates over slavery in the eighteenth century and the effects of colonial settlement in the nineteenth, 'part of the genealogy of a modern British sense of responsibility for the plight of distant strangers' (p. 277). It was a sentiment and a politics which grew up both within and in opposition to the hegemonic imperialist project. It was also a form of universalism which paid little attention to the voices of colonised people themselves. The 'plight' of distant others, though acknowledged to be a result of British action, was none the less tied to their 'backwardness'. The 'distance' of these strangers was thus in both space *and* time: they could not be conceived of as coeval. Many of the varieties of 'telescopic philanthropy' (Robbins, 1990) in those days took a similar form. Gary Bridge (2000) has traced a shift through different ethical systems characterised as liberalindividualist (strong universalism), Habermasian (weak universalism), communitarian (situated) and postmodern (with an emphasis on difference and particularity). In an imaginative move he relates each of these to the conception of space which underlies them: for liberal-individualist it is abstract space; for Habermasian, public space (in that particular version); for communitarian, community/local space; for postmodern, corporeal/intimate space. The shift towards the local is impressive and not encouraging. As Bridge points out, communitarianism tends towards the building of enclosed and excluding spaces, while the postmodern version can resolve into 'a form of passive cosmopolitanism' (p. 527) (the result of a combination of a focus on difference and a hostility towards the traditional action-orientation of Western ethics).

Whatever the routes through which it has arrived, there is a persistent Russian-doll geography of ethics, care and responsibility: from home, to local place, to nation.[22] There is a hegemonic understanding that we care first for, and have our first responsibilities towards, those nearest in. It is a geography of affect which is territorial and emanates from the local. Stanley Cohen's steady-eyed investigation of *States of denial* (2001) asks, 'If there is a meta-rule of looking after your "own people" first, has the threshold for responding to the plight of distant strangers been reached?' (p. 289; see also Bauman, 1993; Geras, 1998). On the one hand, there are arguments that 'the boundaries of "moral impingement" have been widened' (Cohen, 2001, p. 290). On the other, the 'free market of late capitalism – by definition a system that denies its immorality – generates its own cultures of denial' (p. 293) which are buttressed by spatial strategies which include not only distancing but also segregation and exclusion. It may also be, as Bridge and others suggest (see Corbridge, 1993; and Low, 1997), that this Russian-doll ethical imagination has in the West recently become more accentuated. (And yet, the ties of migration, of diasporic communities, even of the networks of cyberspace people-like-us and the different degrees of empathy, bearing no relation to physical planetary distance, which world events evoke, are immediately disruptive of this geography, dislocating any automaticity of relation between social and physical distance and indicating the potential for further change.) Yet the dominant geography is in parts of academe reflected in and exacerbated by an absorption with interiorised temporalities, by a focus on hybridity-at-home in Western cities at the expense of multiplicities elsewhere (Spivak, 1990) and by the persistent opposition of place-as-real to space-as-abstract. In an age when the grotesque realities of the relations of global space are so pressing, this is peculiarly ironic. There is, in these terms, a localisation of ethical commitment at the very moment of increasingly geographically expansive interconnectedness. It raises the question of whether, in a relational and globalised spatiality, 'groundedness', and the search for a situated ethics, must remain tied to notions of the local. If places pose, in highly variable form, the question of our living together in the sense of juxtaposition (throwntogetherness), there is also the question of the negotiation of those, equally varied, wider relations within which they are constituted.

This is already a huge and hotly contested area (Benhabib, 1992; Nussbaum, 1996; Robbins, 1999). It might be, however, that being more explicit about the spatialities which the various contestants bring to the arena would clarify – and shift – some of the terms of debate. One element which is persistent is the territorial character of the different mappings of emotion, loyalty and potential ethical positions. Often what seems to be at issue is merely the size of the relevant territory – a shift of loyalty and identification from one territorial enclosure to a bigger one. Bryan Turner, in his consideration of 'cosmopolitan virtue, globalization and patriotism', is explicit about this:

> The weakness of socialist internationalism was that it had difficulty creating a sense of solidarity without place. The geography of emotions therefore appears to be important in creating civic loyalties and commitment…. Without such a geographical sense of place, republicanism would commit the same mistake as 19th-century socialist internationalism. It would be devoid of emotional specificity'. (2002, p. 49)

The question I want to pose to this is: does it have to be place? Does it have to be territorial at all? Perhaps it is not 'place' that is missing, but grounded, practised, connectedness. (The negotiations of place of Chapters 12 and 13 do not create bounded territories but constellations of connections with strands reaching out beyond.) Turner's own exemplification of trade centres in the ancient world in a sense confirms this – what was crucial was connection. In a globalised world, that kind of connectedness, a practised interrelation, is not confined within place. Thus Corbridge's 'reluctance to substitute a poetics of fragmentation [elsewhere he calls it a poetics of place – p. 460] for the sins of the metanarrative' (1993, p. 460) is well taken, but maybe these aren't the only options. Recognising the open and relational construction of the local enables not a *poetics* of place (as Corbridge argues, this is one of the options being pressed upon us) but a politics of grounded connectedness. If, consistently with a relational space, and abandoning the oppositions of place and space, a relational ethics (Whatmore, 1997) is adopted, very different geographies of affect and of loyalty become possible to imagine.

Moira Gatens and Genevieve Lloyd, in their absorbing interpretation of Spinoza (*Collective imaginings*, 1999), draw out a politics of relatedness which enables a reimagination of the notion of responsibility ('Spinozistic responsibility' they call it). Crucial to their argument is the idea of 'a basic sociability which is inseparable from the understanding of human individuality' (p. 14) (see Chapter 5 above). They link up with Etienne Balibar's concept of 'transindividuality': it is 'impossible strictly speaking to have a strong notion of singularity without *at the same time* having a notion of the interaction and interdependence of individuals' (Balibar, 1997, pp. 9–10, n. 9, cited in Gatens and Lloyd, pp. 121–2; emphasis in the original), and also with Deleuzian workings with the concept of ethology.

Moreover, this inseparability of individuality and interdependence is drawn by Gatens and Lloyd through Spinoza's concept of imagination which they interpret as connected but not limited to the cognitive. It has affective dimensions and this in turn lends it a corporeality. As Gatens and Lloyd put it at one point: 'For him [Spinoza], … imagination involves awareness of other bodies at the same time as our own' (1999, p. 23). This is already very different from that self-absorbed (attempt at) self-constitution which was linked in Chapter 5 with the prioritisation of a (particular understanding of) Time. If, however, 'experience' is not an internalised succession of sensations but rather consists of 'a teeming multiplicity of things and relations that constantly associate and interact' (Hayden, 1998, p. 89, writing of Deleuze), then its spatiality is as significant as its temporal dimension. In an astute aside and with specific reference to academe, Grossberg points out that 'thinking in terms of space demands that intellectuals think of themselves in relation to others in a way that temporal thinking does not permit' (1996, p. 187, n. 19).[23] For Gatens and Lloyd, this awareness of others is predicated on positivity, and is a philosophy of affirmation: 'There is … an inherent orientation of joy towards engagement with what lies beyond the self, and hence towards sociability; and there is a corresponding orientation of sadness towards disengagement and isolation' (1999, p. 53). The consequent question concerns the *nature* of the engagement.

There are many ways in which this approach links to the argument here. *First*, there are parallels. A full recognition of the characteristics of space also entails the positive interconnectivity, the nature of the constitutive relationality, of this approach. And as Gatens and Lloyd stress, along with Balibar, this is a relational ontology which avoids the pitfalls both of classical individualism and of communitarian organicism; just so a full recognition of space involves the rejection both of any notion of authentic self-constituting territories/places and of the closed connectivities of structuralism as spatial (and thus evokes space as always relational and always open, being made) and implies the same structure of the possibility of politics.[24] It picks up on the positive concepts of space in those strands of philosophy explored in Part *Two* – Bergson's multiple durations, Laclau's event – and leaves behind those other uses of the term, within the same philosophies, which so restrict an appreciation of the liveliness of space.

But this is more than a matter of parallels. The *second* claim I want to make is that this approach to the understanding of the social, the individual, the political, itself *implies and requires* both a strong dimension of spatiality and the conceptualisation of that spatiality in a particular way. At one level this is to rehearse again the fact that any notion of sociability, in its sparest form simply multiplicity, is to imply a dimension of spatiality. This is obvious, but since it usually remains implicit (if even that), its implications are rarely drawn out. The very acknowledgement of our constitutive interrelatedness implies a spatiality; and that in turn implies that the nature of that spatiality should be a crucial avenue of enquiry and political engagement. Further, this kind of interconnectedness which stresses the imaginative awareness of others, evokes the outwardlookingness of a spatial imagination which was explored in Chapter 5. In other words, to push the point further, the full recognition of contemporaneity implies a spatiality which is a multiplicity of stories-so-far. Space *as* coeval becomings. Or again, an understanding of the social and the political which avoids both classical individualism and communitarian organicism absolutely requires its constitution through a spatio-temporality which is open, through an open-ended temporality which itself necessarily requires a spatiality that is both multiple and not closed, one which is always in the process of construction. Any politics which acknowledges the openness of the future (otherwise there could be no realm of the political) entails a radically open time-space, a space which is always being made.

There are parallels in modes of argumentation, then. And implications (usually implicit) within political philosophies for the conceptualisation of space. But there is then a *third* realm. If these political philosophies entail a particular way of approaching the conceptualisation of spatiality then they reciprocally raise the question of the spatiality (or spatialities) *of* politics, and the spatialities *of* responsibility, loyalty, care. If we take seriously the relational construction of identity (of ourselves, of the everyday, of places), then what is the potential geography of our politics towards those relations?

London again. The metropolis as a whole and the financial City within it form – as does every place – a distinctive articulation within the power-geometries of today's globalisation. The implacable material presence of the City, in its Square Mile and its newer outposts, defies any imaginary of 'the global' as produced and directed by some force mysteriously located up there or out there. It is here. The built-form attests too to the recognition, through the centuries, that the space in which it deals is more than a matter of overcoming distance; that it also involves endowing the heterogeneity of its multiplicity with heavy symbolic meaning. The City's physical self-assertion in this way contributes also to the hegemonic proposition that globalisation is inevitable in this particular form; a force that can in no way be gainsaid. The financial City is, moreover, the centrepiece of economic strategy for the metropolis and of one version of London's identity.

On this view, certainly, neither the City nor the wider city can be interpreted as local victims of the global. From here run practices of engagement – investment, trading, dealing, disinvestment, exchange, the conjuring of the most fanciful (and variously powerful and disastrously fragile) financial instruments – which extend around the world. A constant interplay with other places, on which it depends, whose future it can make or break. New spaces being made. Here the everyday is indubitably on a planetary scale.

Globalised certainly, but not simply open. As with so many places of global power its widely applauded openness is tightly selective. In the 1990s, in response to IRA bombing, the Square Mile was fiercely cordoned off behind a 'ring of steel'. Anyone trying to pass was checked as an acceptable entrant. There had been bombs elsewhere but only around the City was such a closure enacted. The media documented the queues awaiting entry. And there remains a heavy presence of security. But, unnoticed by the media, centuries of the social constitution of this place, and of the trajectories which entwine here, ensure its ownership, enforce more ordinary closures. Today, as every ordinary day, exclusions are effected (Allen and Pryke, 1994; McDowell, 1997; Pryke, 1991). Yet, in counterpoint, this is not a space where financiers alone may go. The dominant coding hides, but cannot refuse, the entry of cleaners, caterers, the security guards themselves: 'the inability of a dominant space to suppress entirely the diversity and difference within its bounds' (Allen and Pryke, 1994, p. 466). And this intrusion by those who service the City is linked into its own global relations – with family and friends, for instance, in Nigeria, Portugal, Colombia – other globalisations which highlight the particularities of, and the hiatuses and disconnections within, the City's own reach. Yet this place *is* open in the way that matters to the current project of capitalist globalisation. Indeed the very longevity of this form of openness undermines any assertion of the radical newness of globalisation and underlines that what is at issue is not spatial spread. The 'J18' festival of disruption that temporarily shook this place on 18 June 1999 as part of a Global Day of Action was called the 'Carnival against Capital'.

The power and wealth of this place hold out a degree of purchase upon the global relations which spread out from here. And there is in London a relatively progressive city government. The articulation of this place into planetary power-geometries thus poses the question of a politics towards the relations in which it is embedded: not only not a victim but, from a counterglobalisation point of view, a local place more deserving of challenge than defence. There is also no question but that a strong element of the identity of London for many of its residents includes a recognition, even a celebration, of the internal cultural mixity that is part and parcel of its global citydom. This renders even more stark the persistent apparent oblivion of London and Londoners to the external relations, the daily global raiding parties of various sorts, the activity of finance houses and multinational corporations, upon which the very existence of this place depends.

The proposed strategy for London (Greater London Authority, 2001a) is typical in this regard. It understands the city's identity primarily as being a global city and that in turn is defined primarily as a function of the city's position within global financial markets and related sectors. This is presented as an achievement. The strategy offers no critical analysis of the power-relations which must be sustained for this position to be built and reproduced. It does not follow these established relations and current practices out around the world. Its aim indeed is to strengthen even further this financial dominance. It fails to interrogate both London's huge resources and their historical and current mobilisation into power-relations with other places, and the subordination of other places and the global inequalities on which this metropolis depends and upon which so much of its wealth and status have been built. Indeed, when it does turn to address 'relations with elsewhere' the analysis is pervaded by anxiety about competition. This form of self-positioning represents a significant imaginative failure which closes down the possibility of inventing an alternative local politics that might begin to address the wider geographies of the construction of this place.

In none of this is London in the slightest degree unusual. What it involves, however, is the ongoing forging of London's identity as a dominant place in the production of capitalist globalisation. Members of the city's government have made strong statements about the iniquities of capitalism and have, for instance, criticised an arms fair held within its jurisdiction; but the complicity of the centrepiece of the local economy passes unaddressed.

Gatens and Lloyd write that

> [t]he ongoing forging of identities involves integrating past and present as we move into the indeterminate future; and the determining of identities is at the same time the constitution of new sites of responsibility. The processes of

sympathetic and imaginative identification articulated in Spinoza's treatment of individuality and sociability create new possibilities for responsibility at the same time as they create determinate identities which are, however, inherently open to change. (1999, p. 80)

This is an argument which can contribute to the practised making of the identity of place – a *global* sense of place – and to the construction of a place-based politics which responds to that. Gatens and Lloyd's notion of responsibility is relational (it depends on a notion of identity constructed in relation to others), and embodied (it thus connects with the arguments about not opposing an embodied place to an abstract space). It also implies extension – it is not restricted to the immediate or the local. Their concern is to develop this argument in order to explore ways in which there may validly be said to be collective responsibility for the past (their particular concern is with present-day 'postcolonial' Australia's historical responsibilities to Aboriginal society). They write:

In understanding how our past continues in our present we understand also the demands of responsibility for the past we carry with us, the past in which our identities are formed. We are responsible for the past not because of what we as individuals have done, but because of what we are. (p. 81)

In other words, for Gatens and Lloyd, responsibility indeed has extension; but the dimension of extension which concerns them is the temporal. My question is can this temporal extension be paralleled in the spatial? As 'the past continues in our present' so also is the distant implicated in our 'here'. Identities are relational in ways that are *spatio*-temporal. They are indeed bound up with 'the narratives of the past' (Hall, 1990, p. 225) and made up of resources we 'inherit' (Gilroy, 1997, p. 341), but not only did those pasts themselves have a geography, but the process of identity-construction is 'ongoing' (Gatens and Lloyd) *now*. And it has a global geography. To respond to that geography would be to address the spatial counterpoint to an ethics of hospitality. A politics of outwardlookingness, from place beyond place.

A host of 'local' policies suggest themselves which might address the current articulation of London into the power-geometries of globalisation. They range through from challenging the narrow sectoral focus of the current economic strategy, to support for alternative forms of globalisation (trade union, fair trade, cultural links …), to a politics of consumption, to building alliances (rather than competing) with other places. All address in different ways the geography of current practices through which the city currently sustains itself: challenging some, constructing others that were previously missing. They aim to shift the configuration within which the city is set, and to which it contributes. It would evidently be disingenuous to claim that a bundle of strategies such as these would do much to alter the dynamics of neoliberal globalisation. They *would* make some difference in their own right, but the more important effect would be to stimulate public debate about London's current location and role *within* that globalisation. Indeed, provoking debate should itself be an aim. For, again, this place is not a coherent unity. From the conflicting trajectories within capital to the gulfs between the so-called 'fat cats' and the working class of the Isle of Dogs, Londoners are located in radically contrasting and unequal ways in relation to today's globalisation. This is so not only in terms of the effects of globalisation 'on' them but also in the real texture of their imbrication within it and the complexities which can inhere in that (the poorest people buying the sweat-shop clothes). There will, precisely, be argument. There will be contesting political positions. And that in turn – through, for instance, linking the inequalities within the city to the wider inequalities on which it depends and which it daily sustains – might change the terms of the negotiation within London itself; might enable the city itself to be lived a little differently.

This is but one suggestion, one of many potential dimensions of an alternative wider politics of place. Rather than 'responsibility', Fiona Robinson has explored the, currently restricted but potentially wider, geographies of care. In her book *Globalizing care: ethics, feminist theory, and international relations* (1999) she develops 'a critical ethics of care which integrates the relational ethics of care with a critical account of power-relations, difference, and exclusion in the globalizing world order' (p. 104). By working this way she is evading formalised abstractions; the focus is on practised relations. Her approach implies an abandonment of that unwarranted association of space with the abstract (as opposed to place as real) or of global with universal (as opposed to local as specific). Space as well as place is understood as relational and therefore grounded, real. By working also with a critical account of globalisation, she abandons the tendency to associate care with proximity: 'Care does not, at first sight, seem to respond well to distance' (p. 45). Her insistence, however, is that the relationality of care need not be localised, nor territorialised. It entails recognition (of coevalness) and is learned. As such, she argues, the relations of care can be long-distance too. The argument here, however, is a more general one: for an imaginative self-positioning in the world which opens up to the full recognition of the spatial. Gatens and Lloyd stress the force of embodied imagination in social and political life: that it is constitutive rather than merely reflective of 'the forms of sociability in which we live' (1999, p. 143), how in its various forms it is embedded in institutions and traditions: 'One of the social goods which is constitutive of our very identities is the habitation of an imaginary which enhances our powers of action by providing a ground for our feelings of belonging and our claims to social, political and ethical entitlements' (p. 143).

If '[t]he "inner" multiplicity of cultural identity reflects the "external" multiplicity of the relations between bodies' (p. 81) then perhaps it can follow that relationality into a different geography. Gatens and Lloyd themselves, in a few brief lines (p. 137), point tantalisingly to the possibilities – musing upon how a greater transnational interconnectedness might transform both identities and imaginations. If one could parallel Bergson's proposition of throwing oneself into the past, then maybe this could be one element of throwing oneself into the spatial. Within that reorientation the specifics of response and connectivity can be situated. Indeed, to return to the example above, 'responsibility', like hospitality, in some accounts can be read in terms of a one-way-ness (a kind of hierarchical geography of responsibilities) which itself arrogates unto the 'responsible' figure the superiority of a position of power. Rather, what is perhaps crucial is the more complex issue of implication: it is this which thinking relationally (here the *mutual* constitution of global and local) can bring to the fore.

Gatens and Lloyd's concern is with the past, with the temporal dimension. On that dimension of extension 'responsibility for what we are' can bring its own dangers: be too interwoven with guilt, too easily assuaged by apology. As Lynne Segal has commented, on the current spate of apologies for the past, 'Rituals of remembrance designed to prevent the repetitions of past horrors are usually officially sanctioned only when the distance from immediate responsibility for the acts recalled renders them safe from direct demands for intervention, restitution or retribution' (2001, p. 45). These issues are acute when the dimension of extension is temporal. Gatens and Lloyd themselves are arguing for a practical politics and that practical implication would be far harder to avoid if the dimension was spatial and of the present: the geography of *ongoing* identity construction. In the spatial present what we are is what we do.

Gatens and Lloyd do themselves touch on the spatial. But even they tend to stay within an imagination of places rather than taking on the topography of flows. The focus again is both territorial and on the near rather than the distant. They write: 'the experience of cultural difference is now internal to a culture' (1999, p. 78) and they cite Tully, on whom they draw in their own analysis: 'Cultural diversity is not a phenomenon of exotic and incommensurable others in distant lands and at different stages of historical development, as the old concept of culture makes it appear. No. It is here and now in every society' (Tully, 1995, p. 11, cited in Gatens and Lloyd, 1999, p. 78). Well. Cultural diversity is certainly, in part and increasingly, internal to individual societies; but it *is* implacably also a question of different others in distant lands. It would be a grave myopia were we to ignore that wider geography, to forego that aspect of outwardlookingness in the lived geographical imagination.

Space is as much a challenge as is time. Neither space nor place can provide a haven from the world. If time presents us with the opportunities of change and (as some would see it) the terror of death, then space presents us with the social in the widest sense: the challenge of our constitutive interrelatedness – and thus our collective implication in the outcomes of that interrelatedness; the radical contemporaneity of an ongoing multiplicity of others, human and non-human; and the ongoing and ever-specific project of the practices through which that sociability is to be configured.

notes

Notes to Part *One*

(1) Galleano (1973), p. 17, citing 'Indian informants of Fray Bernardino de Sahagún in the Florentine Codex' (p. 287, n. 6). The sources I have drawn upon for this section are: Soustelle, 1956; Townsend, 1992; Vaillant, 1950; Harley, 1990; Berthon and Robinson, 1991.

(2) There has been a long debate over the nature of these presentiments on the part of the Aztecs. A strong version holds to a notion of prophesy (with Cortés as the returning MesoAmerican deity Quetzalcoatl), but this is now widely questioned. It none the less seems to be the case that the approach of the Spaniards, at that Aztec time and from that Aztec direction, evoked strong historical and geographical associations, and such associations were immensely powerful in the Aztec cosmology.

(3) Galleano, 1973, p. 11.

(4) The Julian calendar was then in force.

(5) And so the question became how to abandon this understanding of 'place' and yet retain an appreciation of specificity, of uniqueness; how to reimagine place (or locality, or region) in a more 'progressive' way. How, in other words, we might engage the 'local', the 'regional', while at the same time insisting on internationalism. It was in this context that I worked towards what I would come to call 'a global sense of place' (Massey, 1991a).

(6) There is a link here back to the first proposition. For many anti-essentialists, the real importance of their position (that of challenging the essential – in the sense of unchanging – nature of identities) is that, precisely, it holds open the possibility of change. As already intimated however, and as will emerge more explicitly later, relational construction only really guarantees the possibility of change when the notion of 'relations' is not confined to that of a closed system.

Notes to Part *Two*

(1) The argument in this sub-section is spelled out in much more detail in Massey, 1992a.

(2) The term 'domesticization' reverberates with the long history of the differential gendering of time (masculine) and space (feminine) – see Massey, 1992a.

(3) In contrast to de Certeau I have opted to use the term trajectory (among others) but with its meaning of an irreversible process. De Certeau (though he is not entirely

consistent in this) tends to stress 'narrative'. In contrast, I have tended not to use that term because of the connotation it can carry of interpreted histories, of discourses. The word 'story', though, is equally ambiguous, and I do also use that. See the discussion in Part *One*. One further point on terminology: time-space and space-time are not distinct concepts; the choice of term in general depends on the emphasis of the argument.

(4)   The distinction most commonly made (though equally frequently disputed) is between analytical and Continental strands of philosophy. It is this distinction that is used by Frodeman (1995), for instance, in his analysis of the relationship between physics and geology.

(5)   It might be noted that this tendency of closed systems to run down might be linked up to Caverero's discussion of the preoccupation of so many theorists of time with death: Prigogine and Stengers write that 'for thermodynamics, time implies degradation and death' (1984, p. 129).

(6)   And Deleuze has spoken of a 'secret link' between these philosophers 'constituted by the critique of negativity, the cultivation of joy, the hatred of interiority, the exteriority of forces and relations, the denunciation of power' (1977, p. 12, cited in Massumi, 1988, p. x).

(7)   See Massey, 1999a, for a detailed consideration of this issue, especially in its relation to questions of time and space.

(8)   Though it is important to remember that a Newtonian physics is still entirely adequate for many practical purposes.

(9)   See also Soja, 1996. For the beginnings of a critique of this idea see Massey, 1991c.

(10) These points are linked. Lévi-Strauss set up his kinship systems as binary symmetries between the parts of which, he postulated, there would be balanced exchange. The 'problem' for such a system is impending inertia. (It is at this point that the need for the third term is elaborated.) Lévi-Strauss interprets this – logically necessary but empirically unlikely – rundown into entropy-maximisation as being the result of the *symmetry* of his initial system. I would argue, however, that it might be better, and more generally, specified as a problem of *closure*. With closure there will indeed be a rundown of the system. (What Lévi-Strauss could have done with at this juncture were today's notions of open, dissipative, non-equilibrium, systems.) And the problem of closure in the synchronies of structuralism leads on to the next point in the main text.

Lévi-Strauss recognised this character of the anthropology he was building. Prigogine and Stengers (1984) write that 'structural anthropology privileges those aspects of society where the tools of logic and finite mathematics can be used … . Discrete elements are counted and combined …' (p. 205). (It is a long way from the world of openendedness and probability about which they themselves are writing.) Lévi-Strauss himself pointed to this and contrasted his anthropology with sociology.

(11) The critical characteristic of synchrony in Louis Althusser's critique, and which disables any adequate concept of history, is its internal closure. Althusser characterises the essential section as being *both* an instant (the vertical break, the slice through time) *and* a closed system. It is this double nature, however, which distinguishes it from the structuralists' synchrony, which is characterised only by the latter.

(12) There is a curious (or perhaps not so curious) side-light to be thrown here on a pitched battle which was waged in geographical circles over critical realism.

Critical realism distinguishes between necessity and contingency in its elaboration of explanations and was adopted by some as a means of addressing uniqueness (Sayer, 1984). War was immediately declared. Some 'Marxists', and a good number of others, sneered at the 'reduction' of causes to the status of 'mere contingency'. Contingency was interpreted by them as being far less satisfactory, as an understanding of a state of affairs, than was 'necessity'. In fact, of course, although their derision was poured out in the name of politics, the assumption that everything happens by necessity leaves precious little room for intervention. But this was anyway a misunderstanding of the meaning of 'contingency'. 'Contingent' in critical realism simply means not within the chain of causality currently under investigation. A contingency occurs when a number of such lines interact in some way to affect each other. All may be lines of 'necessity' in themselves. It is their interaction which is contingent. Given this, it is quite wrong to see a 'contingent' influence in an explanation as somehow indicating a subordination of that influence.

(13) Doel is working with a much broader notion of post-structuralism than I am here. My concern at this point is more strictly with a Derridean approach. However, even given this, I still retain this difference in interpretation from Doel.

(14) While I agree with Houdebine in the very specific terms which I have recounted here, I do not concur with his wider position, most particularly his emphasis upon 'dialectical contradiction'.

(15) The emphasis on specificity is important for the argument here. Part of the argument about 'places', for instance, is that they are not entities of the same order as, say, living organisms: the play between internal and external relations is quite different.

Notes to Part *Three*

(1) 'The political challenge of relational space' was the title of a Vega Day symposium held in Stockholm in April 2003, and published in *Geografiska Annaler* Series B, vol. 86B, no. 1, 2004.

(2) 'Hegemonic' because it was by no means the only understanding of space; and hegemonic only within this sphere. There were other, equally powerful, understandings in other spheres (such as in the relation to representation), some of which coexisted in contradiction.

(3) And not only 'the social' in the sense of the human. In the synthesising studies of regional geography which, classically, defined bounded regions and then recounted them in a sequence from geology to politics, this notion of 'space as already divided up' mapped everything from physical structure to cultural practices. Natter and Jones (1993) aptly refer to it as 'regional geography's paradigmatic narrative strategy' (p. 178). The practice continues today in notions of indigeneity and strategies of geographical containment in relation to the non-human organic world (Whatmore, 1999).

(4) The problem of the reification of 'scale' is a bigger issue, not addressed here – see Amin, 2001.

(5) This is not a 'now' in the sense of a coherence. More generally, there is no implication here that removing the temporal convening of space would *ipso facto*

remove notions of inequality, 'the primitive', etc. Lewin (1993, pp. 133–4) points out that the notion of a 'chain of being' from low to high within the nonhuman organic world is very deep in our culture. Originally, he argues, this was not a story of development through time. Only with Darwin did it become transformed into a story rather than a coexistence of (unequal) difference.

(6) Indeed, it is in these terms – that is, about the existence of other temporalities and stories – that the argument against modernity's dominant formulation is usually posed. Thus, as was seen in the previous chapter, Althusser struggled to conceptualise the possibility of a plurality of times. The existence of such a plurality of trajectories is precisely one of the things which disrupts the possibility of the essential section (a coherent, synchronic 'now').

(7) This was the narrativisation that structuralist anthropology wished to avoid.

(8) 'Allochrony' is the term Fabian uses to capture the denial of coevalness.

(9) I would have a significant reservation here, which is that while this establishment of coevalness, or subjecthood, might have happened, or at least exist as a recognised challenge, *within* the West, it has surely not been achieved between the West and the majority of the world. Indeed, celebrations of hybridity and arguments over multiculturalism within the Western metropoles have to some extent stood in for, or replaced, an older (and admittedly itself problematical) internationalism.

(10) But Jameson (1991) also refers to representation-as-spatialisation: see pp. 156–7 for instance, and the subsequent discussion. And note, too, that Laclau brought in 'physical space', which was in this sense not spatial at all.

(11) The arguments in Chapters 6 and 8 draw on Massey, 1999c.

(12) Dodgshon (1999) has pointed to some of the contradictions inherent in some much-used terminology, most particularly time-space compression and time-space convergence.

(13) Shield's argument is tied in with a view of space as originarily unstructured (see pp. 189–90). In this he is drawing on Hegel, for whom 'Differentiation enters into pure space only as the negation of the original purity' (Shields, citing Derrida, 1970, on Hegel's 'Philosophy of nature'), and on Deleuze and Guattari, and it establishes a relation between spatialisation and temporalisation. One could, however, take serious issue with the postulation, even at the conceptual level, of an originary smooth space in that sense.

(14) For rather different versions of this see Chakrabarty, 2000, and Kraniauskas, 2001.

(15) The argument here is spelled out more fully in Massey, 2004. The agency that place *is* allowed in such formulations is that of the medium through which differentiation is produced.

Notes to Part *Four*

(1) It is the use of the word 'map' here which is significant. Jameson, 1991, in fact keeps coming back to 'mapping', to cartography, to the 'real' nature of mapping, to whether cognitive maps are 'really' maps.

(2) See John Keats, 'On first looking into Chapman's *Homer*', lines 11–13.

(3) Rabasa also notes that 'The emblem of the geographer as Atlas represents the task of cartography as moving from one stable global totality to another where details are corrected.' It is thus that: 'As such the *Atlas* is a palimpsest' (1993, p. 250, note 21).

(4) This was a position which subsequently generated a fascinating debate which touched on the relation of 'space in general' to the specific space of a building, the role of architects and the nature of chance itself. On the one hand buildings were to leave people free both for chance encounters and to create what they wanted of the space (these two things tended to be elided – perhaps because of the conceptual difficulty, in this period, of really taking 'chance' seriously? – see below and later in the chapter). On the other hand, there were clearly patterns of behaviour which architects might study and enable. Emphasis on one or the other of these was an element in arguments between the more anarchistic COBRA and Team 10. As Sadler writes 'Team 10 was right to pay attention to "patterns of association", a situationist might have argued, but it was wrong to then congeal those patterns into fixed "place-forms". The choices left to the inhabitants of a Team 10 structure, as they scurried along its burrows, had in effect already been made by the designers' (Sadler, 1998, p. 32). It was a clash over the role of the architect: 'the situationists asking architects to renounce their master visions …, Team 10 asking architects to press on until the very fundamentals of habitat had been discovered' (p. 32). But it was also a clash over the nature and reality of chance and specifically the chance of space. Had Team 10 pursued their path to the end there might have been no undecidability left. Van Eyck himself followed the Team 10 route and the work of structuralist anthropology: 'If human "patterns of association" were governed by the basic structure of primordial relations, then so would [be] their container, the architectural place-form' (Sadler, 1998, p. 171).

(5) I am summarising only the lines of Lechte's argument most relevant to the concerns here.

(6) It can be argued that while the long-term reconceptualisation of physics leads from examination of deterministic reversible processes to the recognition of stochastic and irreversible ones, quantum mechanics has achieved only an intermediate stage on this journey. It includes probability but not irreversibility. Prigogine and Stengers (1984) wish to push it on to do so, but others – they say – wish to reclaim classical orthodoxy. See also the argument of Thrift (1999).

(7) For a fuller rumination on the space-time of this journey see Massey, 2000c.

(8) As Rabasa points out (1993, p. 44), de Certeau is conscious that his approach is one with a particular history, and that it has effects (de Certeau, 1988, pp. 211–12).

(9) This quotation continues: 'This objectification enabled appropriation of the territories' (p. 52). Here I would part company with him. Appropriation also required cannons and horses and other material supports. Rabasa's analysis seems to remain within the discursive (see 1993, pp. 224–5, footnote 6).

(10) It is also an argument which very constructively challenges the simplistic formulation which would have it that current tendencies towards a return to place, and towards a defensiveness of the local, are a product only of a reaction to the invasive and disorienting processes of globalisation.

(11) The movement of terminology here is interesting: ideas of complexity, complexity theory, the metaphors of complexity. The instability is indicative of the wider point being made. Thrift 'assume[s] that complexity theory is deeply metaphorical' (1999, p. 36).

(12) The argument here refers to the nonhuman as well as the human. As Sarah Whatmore points out, 'Efforts like the UN Convention on Biological Diversity to fix their place in the world as "indigenous species" within "natural habitats" are a no less political regulation of mobile lives than the paraphernalia of passports and border control' (1999, p. 34). 'Atomistic spaces' for 'nature' too?

(13) Thanks to Christine Marsland for persistent questioning, and long conversations, about all this.

(14) The term is evidently problematical. Not only is the whole division between social (meaning human) and natural both contested and constructed and (perhaps) dubious, but – as I was told severely by an earth scientist, while trying to think through these arguments – 'Europe's landscape has been totally artificial for over 4000 years'; and there is plenty of 'nature' within the city too. The fact of nature-culture reinforces my argument. The spatio-temporal specificity of such attitudes is marked. Clark (2002) shows convincingly how, just as industrialising and urbanising Europe 'grew ever more distant from the flux and volatility of the bio-physical world … an almost inverse experience characterized the temperate periphery, where it was difficult for anyone to fully detach themselves from the "flows of grass, water, herds" and other biomaterial elements' (pp. 116–17).

(15) I am grateful for help with all this to John Thornes (of King's College, London), Jim Rose (of Royal Holloway) and Steve Drury and Nigel Harris from Earth Sciences at the Open University. See also Windley, 1977.

(16) 'Layers'. In previous work I have used the term layers, but it was persistently read as 'a geological metaphor' (see the commentary in Massey 1995c; Postscript in the second edition). On this reading the layers have little temporality and still less mutual interaction – which wasn't what I meant at all. My critique of 'palimpsest' rehearses some of the arguments.

(17) In such a way, being 'right here', 'here and now' *is* the encounter (say) rather than the encounter 'taking place' here and now. There are resonances here of Heidegger's co-conceptualisation of entity and placing. As Elden (2001) points out, Heidegger came to argue that we must 'learn to recognise that things in themselves are places, and not only occupy a place' (cited in Elden, p. 90). This was one aspect of Heidegger's struggle to conceive of space in a manner resolutely non-Cartesian; to get away from an imagination of space as extension where that implies an external geometric. It was a reconceptualisation famously integral to the 'turn' in Heidegger's work. But Elden argues that the turn involved also a second move, and this seems more problematical. Elden's argument here is that, having shifted from his earlier prioritisation of time over space, Heidegger first opposed space and place but then moved to reconceptualise space *as* place. In the earlier formulation space was set apart as the sphere of the abstract geometry of extension, and both opposed to place and rejected. In the later work, space itself came to be thought of through its relation to place(s). Although in principle perhaps it need not, this manner of the placing of space both makes it more difficult to imagine space as relational (relations between distant places, Castells' space of flows, today's spaces of globalisation) and works against an understanding of place itself (*Ort*) as open, porous, on the move; a meeting of trajectories.

(18) In fact, one of the conceptualisations of place which they cite in exemplification of this point is my own (in Massey, 1991a, 'A global sense of place'). I think there may have been some misunderstanding here: at any rate we would seem to be in agreement over the internal disjunctive multiplicities of place.

(19) For an exploration of these lines of enquiry see Massey, Quintas and Wield, 1992.

(20) On the 'pure form' of UK Science Park full-scale production was explicitly forbidden. On spatial divisions of labour see Massey, 1995c.

(21) For a more detailed attempt to spatialise Noble's account see Massey, 1997b.

(22) The Lucas Aerospace workers' alternative plan drew upon innovative ideas of both tacit knowledges and alternative products (see Wainwright and Elliott, 1982).

Notes to Part *Five*

(1) Many thanks to Jana Häberlein for leading me to this story and talking through its complexities.

(2) And how often that is true. The church spire or tower, of biscuit-tin and John Major fame in its incarnation of Englishness, celebrates a religion with routes to the West Bank. 'The state bird of Hawaii, the Hawaiian goose, or nene, … evolved from chance arrivals of Canadian geese….' (Williams, 2000, p. 39). And so on.

(3) Donald follows Laclau in writing of a 'return', but there was no originating moment. There is something also in that differentiation between objective presence and contingency which mirrors the imagined opposition between space and time. The infusion of space with time, which is also part of Donald's project (see his pages 139 et seq., and 123), is also a constant reminder of that contingency.

(4) This is, of course, to challenge that old association of community and place – the oft-hailed 'local community'. It is a term which is wielded as an evocation (one might even say an invocation) in many a political and planning document (the UK's New Labour is expert at it).

(5) This line of thinking has been developed in Open University, 1999.

(6) What follows is, inevitably, a very broad-brush picture. For some of the crucial documents in the debate see Greater London Authority, 2001a, 2001b and 2002. The question of how to define the 'world-cityness of London' was a central stake in the political discussion – see below.

(7) One element only; the claim is not that it is the sole cause. Public sector wages and macro-economic strategy also contribute. So does the in-migration which refills the ranks of the poor – in part due to London's attractiveness as a world city.

(8) This discussion of politics for London draws on my own involvement in the process (see, for instance, Massey, 2001b). In one session, when I put it to New Labour representatives that they might have to choose between the City and the poor they simply refused. This is the 'politics without adversaries' discussed by Chantal Mouffe (1998). See also the documents cited in note 6. The Scrutiny Document (Greater London Authority, 2002) is exceptional in trying to get to grips with this issue.

(9) A part of this history has been documented in somewhat greater detail in Massey, 1992b.

(10) By no means all the resentments over this period, however, concerned ethnicity. Holtam and Mayo (1998) recall the housing allocation system at Timber Wharves generating resentment among people in work who could not afford the rent but who saw others 'on benefit' being able to move in.

(11) Nick Jeffrey (1999) has written about a similarly acute situation in south London.

(12) There are a number of things in this argument that might raise an eyebrow – some of them are discussed later in the chapter.

(13) It is not possible to do justice here to the complexity of this politics, nor to its evolution over time. For one indication of this see Bové and Dufour, 2001, including the '10 principles' set out as Appendix 2.

(14) There is, of course, another issue here: that the rejection of US influence could derive from 'French food snobbery'. Naomi Klein, in her Foreword to Bové and Dufour, rejects this too and the protests themselves are confined at least in their explicit targets to such questions as health, quality and diversity.

(15) Robin Cook famously made this statement.

(16) Bové's acknowledged roots within the Left run from Bakunin through the experiences of the Jura Federation. There is also a relation to the formulations of Hardt and Negri (2001) (the use of the term 'multitude', for instance). Yet there is in Bové and Dufour's grounded politics a clear awareness of, and attention to, the existence of different constituencies, different struggles, of the need to negotiate between them and of the practical difficulties of doing so.

(17) Although the persistent tenor of their writing is to favour smooth over striated.

(18) Donald is working here with a distinction 'between the intractable *singularity* of home for him [Williams] (or for you or me) as opposed to the *idea* of community' (p. 151). It is a distinction I am wary of, particularly in its universalising assertions/ impositions (that 'we' are all longing for some self-identical Home) and, of course, in light of trenchant feminist critique. But the wider point he is making remains very useful.

(19) The research was funded by the ESRC, grant no. R000233004: 'High-status growth? Aspects of home and work around high-technology sectors', and was carried out at the Open University, as part of a wider 'South East programme' (see Allen, Massey and Cochrane, 1998), and with Nick Henry now at CURDS at the University of Newcastle. Further details of the work can be found in Henry and Massey, 1995 and Massey, 1995b.

(20) Briefly, the bundles of axes around which this dominance seemed to be constructed gathered around the following: (i) the force of the wage relation and of the market, (ii) the status of Mind/Science/Reason in relation to body, home and the everyday, (iii) gender, as effective and reproduced both through the 'masculinity' of the laboratory and the 'femininity' of the home and through the ongoing, daily, unequal relations between already-established genders within the homes.

(21) Dave Featherstone's work (2001) offers a detailed exemplification of this critique and of the alternative. He contrasts Harvey's (1996) use of the notion of militant particularism with his own analyses of a variety of local struggles showing how they were each continually evolving products of wider relations through which their political identities were moulded.

(22) This Russian-doll geography of affect is intimately related to the preoccupation with scale (i.e. size of *territory*) rather than with a recognition of interconnectedness which is a significant current within geography (see for an excellent critique Amin, 2001, and also Sheppard, 2002). Robbins (1990) provides a hopeful engagement with the possibilities of going beyond the nation-state.

(23) Grossberg acknowledges Carol Stabile for pointing this out to him.

(24) This is also, obviously, set within a much wider literature. See, for one consideration of the trend, Watson (1998), who is drawing upon recent developments of Bergson. The avoidance of individualism is an expected outcome of developing Spinoza, the avoidance of organicism is perhaps less so, given some interpretations' stress on holism.

bibliography

Adam, B. (1990) *Time and social theory*. Cambridge: Polity Press.

Allen, J. (2003) *Lost geographies of power*. Oxford: Blackwell.

Allen, J. and Pryke, M. (1994) 'The production of service space', *Environment and Planning D: Society and Space*, vol. 12, pp. 453–75.

Allen, J., Massey, D. and Cochrane, A. (1998) *Rethinking the region*. London: Routledge.

Althusser, L. (1970) 'The object of capital', in L. Althusser and E. Balibar (eds), *Reading Capital*. London: New Left Books, pp. 71–198.

Amin, A. (2001) 'Globalisation: geographical aspects', in N. Smelser and P.B. Baltes (eds), *International encyclopaedia of the social and behavioural sciences*, vol. 9. Amsterdam: Elsevier Science, pp. 6271–7.

Amin, A. (2002) 'Ethnicity and the multicultural city: living with diversity', *Environment and Planning A*, vol. 34, no. 6, pp. 959–80.

Amin, A. (2004) 'Regions unbound: towards a new politics of place', *Geografiska Annaler*, Ser. B, vol. 86B, no. 1, pp. 33–44.

Amin, A., Massey, D. and Thrift, N. (2000) *Cities for the many not the few*. Bristol: Policy Press.

Amin, A., Massey, D. and Thrift, N. (2003) *Decentering the nation: a radical approach to regional inequality*. London: Catalyst.

Appadurai, A. (ed.) (2001) *Globalization*. Durham, NC and London: Duke University Press.

Architectural Design (1988) Deconstruction in architecture, vol. 15, no. 3/4.

Balibar, E. (1997) 'Spinoza: from individuality to transindividuality', *Mededelingen vanwege het Spinozahuis*. Delft: Eburon.

Bammer, A. (1992) *Displacements: cultural identities in question*. Bloomington and Indianapolis: Indiana University Press.

Barnett, C. (1999) 'Deconstructing context: exposing Derrida', *Transactions of the Institute of British Geographers*, vol. 24, no. 3, pp. 277–93.

Baudrillard, J. (1988) *America*. London: Verso.

Bauman, Z. (1993) *Postmodern ethics*. Oxford: Blackwell.

Bauman, Z. (2000) 'Time and space reunited', *Time and Society*, vol. 9, no. 2/3, pp. 171–85.

Benhabib, S. (1992) *Situating the self: gender, community and postmodernism in contemporary ethics*. Cambridge: Polity Press.

Benjamin, A. (1999) *Architectural philosophy*. London: Athlone Press.

Berger, J. (1974) *The look of things*. New York: Viking.

Bergson, H. (1910) *Time and free will*. Muirhead Library of philosophy (authorised translation by F.L. Pogson). London: George Allen and Unwin.

Bergson, H. (1911) *Matter and memory* (trans. N.M. Paul and W.S. Palmer). London: George Allen and Unwin.

Bergson, H. (1911/1975) *Creative evolution* (trans. A. Mitchell). Westport, CT: Greenwood Press.

Bergson, H. (1959) *Oeuvres*. Paris: Presses Universitaires de France (translations cited are from Prigogine, 1997).

Berthon, S. and Robinson, A. (1991) *The shape of the world*. London: George Philip/Granada TV.

Bhabha, H. (1994) *The location of culture*. London: Routledge.

Bingham, N. (1996) 'Object-ions: from technological determinism towards geographies of relations', *Environment and Planning D: Society and Space*, vol. 14, pp. 635–57.

Bloch, E. (1932/1962) 'Ungleichzeitigkeit und Pflicht zu ihrer Dialektik' in *Erbschaft dieser Zeit*. Frankfurt: Suhrkamp.

Boardman, J. (1996) *Classic landforms of the Lake District*. The Geographical Association in conjunction with the British Geomorphological Research Group.

Bohm, D. (1998) *On creativity* (ed. L. Nichol). London: Routledge.

Bondi, L. (1990) 'Feminism, postmodernism and geography: space for women?', *Antipode*, vol. 22, pp. 156–67.

Borges, J.L. (1970) 'The Argentine writer and tradition', in *Labyrinths*. London: Penguin, pp. 211–20.

Boundas, C.V. (1996) 'Deleuze–Bergson: an ontology of the virtual', in P. Patton (ed.), *Deleuze: a critical reader*. Oxford: Blackwell, pp. 80–106.

Bové, J. and Dufour, F. (2001) *The world is not for sale: farmers against junk food* (Bové and Dufour interviewed by Gilles Luneau, translated by Anna de Casparis). London: Verso.

Bridge, G. (2000) 'Rationality, ethics, and space: on situated universalism and the self-interested acknowledgement of "difference"', *Environment and Planning D: Society and Space*, vol. 18, pp. 519–35.

Brown, P. (1989) *The body and society: men, women and sexual renunciation in early Christianity*. London: Faber and Faber (first published 1988 by Columbia University Press, New York).

Callon, M. (1986) 'Some elements of a sociology of translation: domestication of the scallops and the fisherman of St. Brieuc bay', in J. Law (ed.), *Power, action and belief: a new sociology of knowledge*. London: Routledge, pp. 196–232.

Campbell, B. (1993) *Goliath: Britain's dangerous places*. London: Methuen.

Carnap, R. (1937) *The logical syntax of language*. London: Routledge and Kegan Paul.

Carter, E., Donald, J. and Squires, J. (1993) *Space and place: theories of identity and location*. London: Lawrence and Wishart.

Casey, E. (1996) 'How to get from space to place in a fairly short stretch of time', in S. Field and K. Baso (eds), *Senses of place*. Santa Fé: School of American Research, pp. 14–51.

Castells, M. (1996) *The rise of the network society*. Oxford: Blackwell.

Cavarero, A. (1995) *In spite of Plato: a feminist rewriting of ancient philosophy*. Cambridge: Polity Press.

Chakrabarty, D. (2000) *Provincializing Europe: postcolonial thought and historical difference*. Princeton, NJ and Oxford: Princeton University Press.

Cheah, P. (1998) 'Given culture: rethinking cosmopolitical freedom in transnationalism', in P. Cheah and B. Robbins (eds), *Cosmopolitics: thinking and feeling beyond the nation*. Minneapolis and London: University of Minnesota Press, pp. 290–328.

Clark, N. (2002) 'The demon-seed: bioinvasion as the unsettling of environmental cosmopolitanism', *Theory, Culture and Society*, vol. 19, no. 1–2, pp. 102–25.

Cockburn, C. (1998) *The space between us: negotiating gender and national identities in conflict*. London: Zed Books.

Cohen, S. (2001) *States of denial: knowing about atrocities and suffering*. Cambridge: Polity Press.

Comedia (1995) *Park life: urban parks and social renewal*. London: Comedia.

Corbridge, S. (1993) 'Marxisms, modernities, and moralities: development praxis and the claims of distant strangers', *Environment and Planning D: Society and Space*, vol. 11, pp. 449–72.

Corbridge, S. (1998) 'Development ethics: distance, difference, plausibility', *Ethics, Place, Environment*, vol. 1, no. 1, pp. 35–53.

Davis, M. (2000) *Ecology of fear: Los Angeles and the imagination of disaster*. London: Picador.

de Certeau, M. (1984) *The practice of everyday life*. Berkeley, CA: University of California Press.

de Certeau, M. (1988) *The writing of history* (trans. T. Conley). New York: Columbia University Press. (Originally published 1975 as *L'écriture de l'histoire*. Paris: Gallimard.)

de Léry, J. (1578) *Histoire d'un voyage faict en la terre du Brésil*.

Debord, G. (1956/1981) 'Theory of the dérive', in K. Knabb (ed. and trans.), *Situationist International Anthology*. Berkeley, CA: Bureau of Public Secrets, pp. 50–4.

DeLanda, M. (2002) *Intensive science and virtual philosophy*. London: Continuum.

Deleuze, G. (1977) 'I have nothing to admit' (trans. J. Forman), *Semiotext(e), Anti-Oedipus*, vol. 2, no. 3, pp. 111–16.

Deleuze, G. (1988) *Bergsonism* (trans. H. Tomlinson and Barbara Habberjam). New York: Zone Books.

Deleuze, G. (1953/1991) *Empiricism and subjectivity* (trans. C.V. Boundas). New York: Columbia University Press.

Deleuze, G. (1995) *Negotiations: Interviews 1972–1990*. New York: Columbia University Press (the original interview cited here was published in *Libération*, 23 October 1980).

Deleuze, G. and Guattari, F. (1988) *A thousand plateaus*. London: Athlone Press.

Deleuze, G. and Parnet, C. (1987) *Dialogues* (trans. H. Tomlinson and B. Habberjam). London: Athlone Press.

Derrida, J. (1970) '"Ousia and Gramme": a note to a footnote in *Being and Time*', in *Phenomenology in perspective* (trans. E.S. Casey, ed. F.J. Smith). Dordrecht: Martinus Nijhoff.

Derrida, J. (1972/1987) *Positions* (trans. A. Bass). London: Athlone Press.

Derrida, J. (1994) 'The spatial arts: an interview with Jacques Derrida' (P. Brunette, D. Wills), in P. Brunette and D. Wills (eds), *Deconstruction and the visual arts: art, media, architecture*. Cambridge: Cambridge University Press, pp. 9–32.

Derrida, J. (1995) *Points ... Interviews, 1974–1994* (ed. E. Webber, trans. P. Kamuf and others). Stanford, CA: Stanford University Press.

Derrida, J. (1996) 'Remarks on deconstruction and pragmatism', in C. Mouffe (ed.), *Deconstruction and pragmatism*. London: Routledge, pp. 77–88.

Derrida, J. (1997) *Politics of friendship*. London: Verso.

Derrida, J. (2001) *On cosmopolitanism and forgiveness*. London: Routledge.

Deutsche, R. (1996) 'Agoraphobia', in *Evictions: art and spatial politics*. Cambridge, MA: MIT Press.

Dirlik, A. (1998) 'Globalism and the politics of place', *Development*, vol. 41, no. 2, pp. 7–13.

Dirlik, A. (2001) 'Place-based imagination: globalism and the politics of place', in R. Prazniak and A. Dirlik (eds), *Places and politics in an age of globalization*. Lanham, MA: Rowman and Littlefield, pp. 15–51.

Dodge, M. and Kitchin, R. (2001) *Mapping cyberspace*. London: Routledge.

Dodgshon, R. (1999) 'Human geography at the end of time? Some thoughts on the notion of time-space compression', *Environment and Planning D: Society and Space*, vol. 17, pp. 607–20.

Doel, M. (1999) *Poststructuralist geographies: the diabolical art of spatial science*. Edinburgh: Edinburgh University Press.

Donald, J. (1999) *Imagining the modern city*. London: Athlone Press.

Elden, S. (2001) *Mapping the present: Heidegger, Foucault and the project of a spatial history*. London: Continuum.

Elphick, J. (ed.) (1995) *Collins atlas of bird migration*. London: HarperCollins.

Escobar, A. (2001) 'Culture sits in places: reflections on globalism and subaltern strategies of localization', *Political Geography*, vol. 20, pp. 139–74.

Fabian, J. (1983) *Time and the Other: how anthropology makes its object*. New York: Columbia University Press.

Featherstone, D. (2001) 'Spatiality, political identities and the environmentalism of the poor', PhD thesis. Milton Keynes: The Open University.

Featherstone, M., Lash, S. and Robertson, R. (eds) (1994) *Global modernities*. London: Sage.

Ferrier, E. (1990) 'Mapping power: cartography and contemporary cultural theory', *Antithesis*, vol. 4, no. 1, pp. 35–49.

Foucault, M. (1980) 'Questions on geography' in C. Gordon (ed.), *Power/knowledge: selected interviews and other writings, 1972–1977*. London: Harvester Wheatsheaf, pp. 63–77.

Frodeman, R. (1995) 'Geological reasoning: geology as an interpretive and historical science', *Bulletin of the Geological Society of America*, vol. 107, no. 8, pp. 960–8.

Galleano, E. (1973) *Open veins of Latin America*. New York: Monthly Review Press. (Originally published 1971 as *Las venas abiertas de América Latina*. Mexico City: Siglo XXI Editores.)

Gatens, M. and Lloyd, G. (1999) *Collective imaginings: Spinoza, past and present*. London: Routledge.

Gates, B. (1995) *The road ahead*. New York: Viking.

Geras, N. (1998) *The contract of mutual indifference: political philosophy after the holocaust*. London: Verso.

Gibson-Graham, J-K. (1996) *The end of capitalism (as we knew it)*. Oxford: Blackwell.

Gibson-Graham, J-K. (2002) 'Beyond global vs. local: economic politics outside the binary frame', in A. Herod and M.W. Wright (eds), *Geographies of power: placing scale*. Oxford: Blackwell, pp. 25–60.

Giddens, A. (1984) *The constitution of society*. London: Harper and Row.

Giddens, A. (1990) *The consequences of modernity*. Cambridge: Polity Press.

Gilroy, P. (1997) 'Diaspora and the detours of identity', in K. Woodward (ed.), *Identity and difference*. London: Sage, pp. 299–346.

Glancey, J. (1996) 'Exit from the city of destruction', *Independent*, 23 May, p. 20.

Glancey, J. and Brandolini, S. (1999) 'Aldo van Eyck: the urban space man', *Guardian*, 28 January, p. 16.

Gleick, J. (1988) *Chaos*. London: Abacus.

Goodchild, P. (1996) *Deleuze and Guattari: an introduction to the politics of desire*. London: Sage.

Graham, S. (1998) 'The end of geography or the explosion of place? Conceptualising space, place and information technology', *Progress in Human Geography*, vol. 22, no. 2, pp. 165–85.

Greater London Authority (2001a) *Towards the London Plan: initial proposals for the Mayor's Spatial Development Strategy*. London: GLA.

Greater London Authority (2001b) *Economic Development Strategy*. London: GLA.

Greater London Authority (2002) *Spatial development strategy investigative committee: towards the London Plan, Final Report*. London: GLA.

Gross, D. (1981–2) 'Space, time and modern culture', *Telos*, vol. 50, pp. 59–78.

Grossberg, L. (1996) 'The space of culture, the power of space', in I. Chambers and L. Curti (eds), *The post-colonial question*. London: Routledge, pp. 169–88.

Grosz, E. (1995) *Space, time, and perversion: essays on the politics of bodies*. New York and London: Routledge.

Grosz, E. (2001) 'The future of space: toward an architecture of invention', in *Olafur Eliasson: surroundings surrounded: essays on space and science* (ed. P. Weibel). Karlsruhe: ZKM Center for Art and Media and Cambridge, MA: MIT Press, pp. 252–68.

Guattari, F. (1989/2000) *The three ecologies* (trans. I. Pindar and P. Sutton). London; New Brunswick, NJ: Athlone Press. (First published as *Les trois écologies* in 1989 by Editions Galillée, Paris.)

Gupta, A. and Ferguson, J. (1992) 'Beyond "culture": space, identity, and the politics of difference', *Cultural Anthropology*, vol. 7, pp. 6–23.

Hacking, I. (1990) *The taming of chance*. Cambridge: Cambridge University Press.

Hall, S. (1990) 'Cultural identity and diaspora', in J. Rutherford (ed.), *Identity: community, culture, difference*. London: Lawrence and Wishart, pp. 222–37.

Hall, S. (1992) 'The question of cultural identity', in S. Hall, D. Held and A. McGrew (eds), *Modernity and its futures*. Cambridge and Milton Keynes: Polity Press in association with The Open University, pp. 273–325.

Hall, S. (1996) 'When was the "post-colonial"? Thinking at the limit', in I. Chambers and L. Curti (eds), *The post-colonial question*. London: Routledge, pp. 242–60.

Haraway, D. (1991) *Simians, cyborgs, and women*. London: Free Association Books.

Harcourt, W. (2002) 'Place politics and justice: women negotiating globalization', *Development*, Special Issue, vol. 45, no. 1.

Hardt, M. and Negri, A. (2001) *Empire*. Cambridge, MA: Harvard University Press.

Harley, J.B. (1988) 'Maps, knowledge, and power', in D. Cosgrove and S. Daniels (eds), *The iconography of landscape: essays on the symbolic representation, design and use of past environments*. Cambridge: Cambridge University Press, pp. 277–312.

Harley, J.B. (1990) *Maps and the Columbian encounter*. Milwaukee, WI: University of Wisconsin.

Harley, J.B. (1992) 'Deconstructing the map', in T. Barnes and J. Duncan (eds), *Writing worlds: discourse, text and metaphor in the representation of language*. London: Routledge, pp. 231–47.

Harvey, D. (1993) 'Class relations, social justice and the politics of difference', in J. Squires (ed.), *Principled positions: postmodernism and the rediscovery of value*. London: Lawrence and Wishart, pp. 85–120.

Harvey, D. (1996) *Justice, nature and the geography of difference*. Oxford: Blackwell.

Haver, W. (1997) 'Queer research; or, how to practise invention to the brink of intelligibility', in S. Golding (ed.), *The eight technologies of otherness*. London: Routledge, pp. 277–92.

Hayden, P. (1998) *Multiplicity and becoming: the pluralist empiricism of Gilles Deleuze*. New York: Peter Lang.

Hayles, N.K. (1990) *Chaos bound: orderly disorder in contemporary literature and science*. Ithaca, NY: Cornell University Press.

Hayles, N.K. (1999) *How we became posthuman: virtual bodies in cybernetics, literature, and informatics*. Chicago and London: University of Chicago Press.

Henry, N. and Massey, D. (1995) 'Competitive times in high tech', *Geoforum*, vol. 26, no. 1, pp. 49–64.

Hirst, P. and Thompson, G. (1996a) *Globalisation in question*. Cambridge: Polity Press.

Hirst, P. and Thompson, G. (1996b) 'Globalisation: ten frequently asked questions and some surprising answers', *Soundings: a journal of politics and culture*, no. 4, pp. 47–66.

Holtam, N. and Mayo, S. (1998) *Learning from the conflict: reflections on the struggle against the British National Party on the Isle of Dogs, 1993–4*. London: Jubilee Group.

Huggan, G. (1989) 'Decolonizing the map: post-colonialism, post-structuralism and the cartographic connection', *Ariel*, vol. 20, no. 4, pp. 115–31.

Ingold, T. (1993) 'The temporality of landscape', *World Archaeology*, vol. 35, pp. 152–74.

Ingold, T. (1995) 'Building, dwelling, living', in M. Strathern (ed.), *Shifting contexts: transformations in anthropological knowledge*. London: Routledge, pp. 57–80.

Irigaray, L. (1993) *The ethics of sexual difference*. New York: Cornell. (First published 1984 as *Ethique de la différence sexuelle*. Paris: Minuit.).

Jacobs, J. (1961) *The death and life of great American cities*. Harmondsworth: Penguin.

Jakobson, R. (1985) *Selected writings – VI: Early Slavic paths and crossroads* (ed. Stephen Rudy). Paris: Mouton.

James, C.L.R. (1938) *The black Jacobins*. London: Allison and Busby.

Jameson, F. (1991) *Postmodernism, or, the cultural logic of late capitalism*. London: Verso.

Jardine, A. (1985) *Gynesis: configurations of woman and modernity*. Ithaca, NY: Cornell University Press.

Jeffrey, N. (1999) 'The sharp edge of Stephen's city', *Soundings: a journal of politics and culture*, no. 12, pp. 26–42.

Jencks, C. (1973) *Modern movements in architecture*. Harmondsworth: Penguin.

Kamuf, P. (ed.) (1991) *A Derrida reader: between the blinds*. New York: Columbia University Press.

Kaplan, C. (1996) *Questions of travel: postmodern discourses of displacement*. Durham, NC and London: Duke University Press.

Katz, C. (1996) 'Towards minor theory', *Environment and Planning D: Society and Space*, vol. 14, pp. 487–99.

Kern, S. (1983) *The culture of time and space, 1880–1918*. Cambridge, MA: Harvard University Press.

King, A. (1995) 'The Times and Spaces of modernity (or who needs postmodernism?)', in M. Featherstone, S. Lash and R. Robertson (eds), *Global modernities*. London: Sage, pp. 108–23.

King, A. (2000) 'Postcolonialism, representation and the city', in S. Watson and G. Bridge (eds), *A companion to the city*. Oxford: Blackwell, pp. 260–9.

Kitchin, R.M. (1998) 'Towards geographies of cyberspace', *Progress in Human Geography*, vol. 22, no. 3, pp. 385–406.

Kraniauskas, J. (2001) 'What will be: review of Chakrabarty, 2000, and Harootunian, 2000', *Radical Philosophy*, no. 107, pp. 43–5.

Kroeber, K. (1994) *Ecological literary criticism: romantic imagining and the biology of mind*. New York: Columbia University Press.

Laclau, E. (1990) *New reflections on the revolution of our time*. London: Verso.

Laclau, E. and Mouffe, C. (2001) *Hegemony and socialist strategy*, 2nd edn. London: Verso. (First published 1985; page references are to the 2001 edition.)

Lapham, L. (1998) *The agony of mammon: the imperial global economy explains itself to the membership in Davos, Switzerland*. London: Verso.

Latour, B. (1993) *We have never been modern* (trans. C. Porter). London: Harvester Wheatsheaf. (Originally published 1991 as *Nous n'avons jamais été modernes*. Paris: Editions La Découverte; page references are to the 1993 edition.)

Latour, B. (1999a) *'Ein Ding ist ein Thing* – a philosophical platform for a left (European) party', *Soundings: a journal of politics and culture*, no. 12, Summer, pp. 12–25.

Latour, B. (1999b) *Pandora's hope: essays on the reality of science studies*. Cambridge, MA and London: Harvard University Press.

Latour, B. (2004) *Politics of nature: how to bring the sciences into democracy* (trans. C. Porter). Cambridge, MA: Harvard University Press.

Lechte, J. (1994) *Fifty key contemporary thinkers: from structuralism to postmodernity*. London: Routledge.

Lechte, J. (1995) '(Not) belonging in postmodern space', in S. Watson and K. Gibson (eds), *Postmodern cities and spaces*. Oxford: Blackwell, pp. 99–111.

Leech, K. (2001) *Through our long exile: contextual theology and the urban experience*. London: Darton, Longman and Todd.

Lefebvre, H. (1991) *The production of space* (trans. D. Nicholson-Smith). Oxford: Blackwell.

Lester, A. (2002) 'Obtaining the "due observance of justice": the geographies of colonial humanitarianism', *Environment and Planning D: Society and Space*, vol. 20, pp. 277–93.

Levin, Y. (1989) 'Dismantling the spectacle: the cinema of Guy Debord', in E. Sussman (ed.), *On the passage of a few people through a rather brief moment in time: the Situationist International 1957–1972*. Cambridge MA: MIT Press, pp. 72–123.

Lévi-Strauss, C. (1945/1972) 'Structural analysis in linguistics and anthropology', in *Structural Anthropology* (trans. Claire Jacobson and Brooke Grundfest Schoepf). Harmondsworth: Penguin, pp. 31–54.

Lévi-Strauss, C. (1956/1972) 'Do dual organizations exist?', in *Structural Anthropology* (trans. Claire Jacobson and Brooke Grundfest Schoepf). Harmondsworth: Penguin, pp. 132–63.

Lewin, R. (1993) *Complexity: life at the edge of chaos*. London: J.M. Dent.

Little, P. (1998) 'Globalization and the struggles over places in the Amazon', *Development*, vol. 41, no. 2, pp. 70–5.

Lloyd, G. (1996) *Spinoza and the Ethics*. London: Routledge.

Low, M. (1997) 'Representation unbound: globalization and democracy', in K. Cox (ed.), *Spaces of globalization: reasserting the power of the local*. London: Guilford Press, pp. 240–80.

Low, M. and Barnett, C. (2000) 'After globalisation', *Environment and Planning D: Society and Space*, vol. 18, pp. 53–61.

Lyotard, J-F. (1989) 'The sublime and the avant-garde', in A. Benjamin (ed.), *The Lyotard reader*. Oxford: Blackwell, pp. 196–211.

MacEwan, A. (1999) *Neoliberalism or Democracy? Economic strategy, markets, and alternatives for the 21st century*. London: Zed Books.

Macpherson, J.G. (1901) 'Geology', in *A history of Cumberland, volume 1* (ed. J. Wilson) (in *The Victoria History of the Counties of England*, edited by H. Arthur Doubleday). Westminster: Archibald Constable and Company.

Massey, D. (1991a) 'A global sense of place', *Marxism Today*, June, pp. 24–9. (Reprinted in Massey, D. (1994) *Space, place and gender*. Cambridge: Polity Press, pp. 146–56.)

Massey, D. (1991b) 'The political place of locality studies', *Environment and Planning A*, vol. 23, pp. 267–81. (Reprinted in Massey, D. (1994) *Space, place and gender*. Cambridge: Polity Press, pp. 157–73.)

Massey, D. (1991c) 'Flexible sexism', *Environment and Planning D: Society and Space*, vol. 9, pp. 31–57. (Reprinted in Massey, D. (1994) *Space, place and gender*. Cambridge: Polity Press, pp. 212–48.)

Massey, D. (1992a) 'Politics and space-time', *New Left Review*, no. 196, pp. 65–84.

Massey, D. (1992b) 'Double articulation: a place in the world', in A. Bammer (ed.), *Displacements: cultural identities in question*. Bloomington and Indianapolis: Indiana University Press, pp. 110–21.

Massey, D. (1995a) 'Thinking radical democracy spatially', *Environment and Planning D: Society and Space*, vol. 13, pp. 283–8.

Massey, D. (1995b) 'Masculinity, dualisms and high technology', *Transactions of the Institute of British Geographers*, vol. 20, pp. 487–99.

Massey, D. (1995c) *Spatial divisions of labour: social structures and the geography of production*, 2nd edn. Basingstoke: Macmillan. First edn. 1984.

Massey, D. (1996a) 'Politicising space and place', *Scottish Geographical Magazine*, vol. 112, no. 2, pp. 117–23.

Massey, D. (1996b) 'Space/power, identity/difference: tensions in the city', in A. Merrifield and E. Swyngedouw (eds), *The urbanization of injustice*. London: Lawrence and Wishart, pp. 100–16.

Massey, D. (1997a) 'Spatial disruptions', in S. Golding (ed.), *The eight technologies of otherness*. London: Routledge, pp. 218–25.

Massey, D. (1997b) 'Economic: non-economic', in R. Lee and J. Wills (eds), *Geographies of economies*. London: Edward Arnold, pp. 27–36.

Massey, D. (1999a) 'Space-time, "science" and the relationship between physical geography and human geography', *Transactions of the Institute of British Geographers*, vol. 24, pp. 261–76.

Massey, D. (1999b) 'Negotiating disciplinary boundaries', *Current Sociology*, vol. 47, no. 4, pp. 5–12.

Massey, D. (1999c) 'Imagining globalisation: power-geometries of time-space', in A. Brah, M. Hickman and M. Mac an Ghaill (eds), *Future worlds: migration, environment and globalization*. Basingstoke: Macmillan, pp. 27–44.

Massey, D. (2000a) 'The geography of power', in B. Gunnell and D. Timms (eds), *After Seattle: globalisation and its discontents*. London: Catalyst.

Massey, D. (2000b) 'The geography of power', *Red Pepper*, July, pp. 18–21.

Massey, D. (2000c) 'Travelling thoughts', in P. Gilroy, L. Grossberg and A. McRobbie (eds), *Without guarantees: in honour of Stuart Hall*. London: Lawrence and Wishart, pp. 195–215.

Massey, D. (2001a) 'Living in Wythenshawe', in I. Borden, J. Kerr, J. Rendell and A. Pivaro (eds), *The unknown city: contesting architecture and social space*. Cambridge, MA: MIT Press, pp. 458–75.

Massey, D. (2001b) 'Opportunities for a world city: reflections on the draft economic development and regeneration strategy for London', *City*, vol. 5, no. 1, pp. 101–5.

Massey, D. (2004) 'Geographies of responsibility', *Geografiska Annaler*, Ser B, vol. 86B, no. 1, pp. 5–18.

Massey, D., Quintas, P. and Wield, D. (1992) *High-tech fantasies*. London: Routledge.

Massumi, B. (1988) 'Translator's Foreward' to G. Deleuze and F. Guattari, *A thousand plateaus*. London: Athlone Press, pp. ix–xv.

Massumi, B. (1992) *A user's guide to capitalism and schizophrenia: deviations from Deleuze and Guattari*. Cambridge, MA: MIT Press.

Mazis, G.A. (1999) 'Chaos theory and Merleau-Ponty's ontology', in D. Olkowski and J. Morley (eds), *Merleau-Ponty: interiority and exteriority, psychic life and the world*. Albany, NY: State University of New York, pp. 219–41.

McClintock, A. (1995) *Imperial leather: race, gender and sexuality in the colonial contest*. London: Routledge.

McDowell, L. (1997) *Capital culture: gender at work in the City*. Oxford: Blackwell.

Merleau-Ponty, M. (1962) *Phenomenology of perception* (trans. Colin Smith). New York: Humanities.

Miller, C.L. (1993) 'The postidentitarian predicament in the footnotes of *A thousand plateaus*: nomadology, anthropology, and authority', *Diacritics*, vol. 23, no. 3, pp. 6–35.

Mitchell, W. (1995) *City of bits: space, place and the infobahn*. Cambridge, MA: MIT Press.

Moore, S. (1988) 'Getting a bit of the other – the pimps of postmodernism', in R. Chapman and J. Rutherford (eds), *Male order: unwrapping masculinity*. London: Lawrence and Wishart, pp. 165–92.

Morris, M. (1992a) 'Great moments in social climbing: King Kong and the human fly', in B. Colomina (ed.), *Sexuality and space*. New York: Princeton Papers on Architecture, Princeton Architectural Press, pp. 1–51.

Morris, M. (1992b) *Ecstasy and economics: American essays for John Forbes*. Sydney: EMPress.

Mouffe, C. (1991) 'Citizenship and political community', in Miami Theory Collective (eds), *Community at loose ends*. Minneapolis: University of Minnesota Press.

Mouffe, C. (1993) *The return of the political*. London: Verso.

Mouffe, C. (1995) 'Post-Marxism: democracy and identity', *Environment and Planning D: Society and Space*, vol. 13, pp. 259–65.

Mouffe, C. (1998) 'The radical centre: a politics without adversary', *Soundings: a journal of politics and culture*, no. 9, Summer, pp. 11–23.

Nancy, J-L. (1991) *The inoperative community*. Minneapolis: University of Minnesota Press.

Nash, C. (2002) 'Genealogical identities', *Environment and Planning D: Society and Space*, vol. 20, pp. 27–52.

Natter, W. and Jones, J-P. (1993) 'Signposts towards a post-structuralist geography', in J.P. Jones, W. Natter and T.R. Schatzky (eds), *Postmodern contentions: epochs, politics, space*. New York: Guilford Press, pp. 165–203.

Negroponte, N. (1995) *Being digital*. London: Hodder and Stoughton.

Noble, D.F. (1992) *A world without women: the Christian clerical culture of Western science*. Oxford: Oxford University Press.

Nussbaum, C. (1996) *For love of country: debating the limits of patriotism* (ed. J. Cohen). Boston, MA: Beacon Press.

Oakes, T.S. (1993) 'Ethnic tourism and place identity in China', *Environment and Planning D: Society and Space*, vol. 11, pp. 47–66.

Ohmae, K. (1994) *The borderless world: power and strategy in the interlinked economy*. London: HarperCollins.

Ondaatje, M. (1992) *The English patient*. New York: Vintage/Random House.

Open University (1997) *Earth and life* (four volumes). Milton Keynes: The Open University.

Open University (1999) *Understanding cities* (three volumes). London and Milton Keynes: Routledge in association with The Open University.

Osborne, P. (1995) *The politics of time: modernity and avant garde*. London: Verso.

Patton, P. (2000) *Deleuze and the political*. London: Routledge.

Peet, R. (2001) 'Neoliberalism or democratic development? Review of MacEwan, 1999', *Review of International Political Economy*, vol. 8, no. 2, pp. 329–43.

Pellerin, H. (1999) 'The cart before the horse? The coordination of migration policies in the Americas and the neoliberal economic project of integration', *Review of International Political Economy*, vol. 6, no. 4, pp. 468–93.

Pinder, D. (1994) 'Cognitive mapping: cultural politics from the situationists to Fredric Jameson', paper presented at the 'Mapping and transgressing space and place' session of the Annual Association of American Geographers, San Francisco, April.

Plato (1977) 'Timaeus', in *Timaeus and Critias* (trans. D. Lee). Harmondsworth: Penguin.

Pratt, A. (2000) 'New media, the new economy and new spaces', *Geoforum*, vol. 31, pp. 425–36.

Pratt, G. (1999) 'Geographies of identity and difference: marking boundaries' in D. Massey, J. Allen and P. Sarre (eds), *Human geography today*. Cambridge: Polity Press. pp. 151–67.

Pratt, G. and Hansen, S. (1994) 'Geography and the construction of difference', *Gender, place and culture*, vol. 1, no. 1, pp. 5–29.

Prigogine, I. (1997) *The end of certainty: time, chaos and the laws of nature*. London: Free Press.

Prigogine, I. and Stengers, I. (1984) *Order out of chaos*. London: Heinemann.

Pryke, M. (1991) 'An international city going "global": spatial change in the City of London', *Environment and Planning D: Society and Space*, vol. 9, pp. 197–222.

Rabasa, J. (1993) *Inventing America: Spanish historiography and the formation of Eurocentrism*. Norman, OK and London: Oklahoma University Press.

Raffles, H. (2002) *In Amazonia: a natural history*. Princeton, NJ and Oxford: Princeton University Press.

Rajchman, J. (1991) *Truth and eros: Foucault, Lacan and the question of ethics*. London: Routledge.

Rajchman, J. (1998) *Constructions*. Cambridge, MA: MIT Press.

Rajchman, J. (2001) 'Thinking the city', paper delivered to Thinking the City conference, Tate Modern and ESRC, April, mimeo.

Robbins, B. (1990) 'Telescopic philanthropy: professionalism and responsibility in *Bleak House*', in H. Bhabha (ed.), *Nation and narration*. London: Methuen, pp. 213–30.

Robbins, B. (1999) *Feeling global: internationalism in distress*. New York and London: New York University Press.

Robins, K. (1997) 'The new communications geography and the politics of optimism', *Soundings: a journal of politics and culture*, no. 5, pp. 191–202.

Robinson, F. (1999) *Globalizing care: ethics, feminist theory, and international relations*. Boulder, CO: Westview Press.

Rodowick, D. (1997) *Gilles Deleuze's time machine*. Durham, NC: Duke University Press.

Rorty, R. (1979) *Philosophy and the mirror of nature*. Princeton, NJ: Princeton University Press.

Rose, G. (1993) *Feminism and geography: the limits of geographical knowledge*. Cambridge: Polity Press.

Rose, S. (1997) *Lifelines: biology, freedom, determinism*. Harmondsworth: Penguin.

Ross, K. (1996) 'Streetwise: the French invention of everyday life', *Parallax* #2, February, pp. 67–75.

Sadler, S. (1998) *The situationist city*. London: MIT Press.

Sakai, N. (1989) 'Modernity and its critique: the problem of universalism and particularism', in Mosao Miyoshi and H.D. Harootunian (eds), *Postmodernism and Japan*. Durham, NC: Duke University Press.

Sartre, J-P. (1981) *La Nausée*, in *Oeuvres romanesques*. Paris: Gallimard.

Sassen, S. (2001) 'Spatialities and temporalities of the global: elements for a theorization', in A. Appadurai (ed.), *Globalization*. Durham, NC and London: Duke University Press, pp. 260–78.

Sayer, A. (1984) *Method in social science: a realist approach*. London: Hutchinson.

Segal, L. (2001) 'Defensive functioning: review of Cohen, 2000', *Radical Philosophy*, vol. 108, pp. 45–6.

Seidler, V.J. (1994) 'Men, heterosexualities and emotional life', in S. Pile and N. Thrift (eds), *Mapping the subject: geographies of cultural transformation*. London: Routledge.

Sennett, R. (1970) *The uses of disorder*. Harmondsworth: Penguin.

Serres, M. (1982) 'Turner translates Cournot' (trans. M. Shortland), *Block*, vol. 6, no. 54, pp. 46–55.

Sharp, J., Routledge, P., Philo, C. and Paddison, R. (eds) (2000) *Entanglements of power: geographies of domination/resistance*. London: Routledge.

Sheppard, E. (2002) 'The spaces and times of globalization', *Economic Geography*, vol. 78, 307–30.

Shields, R. (1992) 'A truant proximity: presence and absence in the space of modernity', *Environment and Planning D: Society and Space*, vol. 10, pp. 181–98.

Sibley, D. (1995) *Geographies of exclusion: society and difference in the West*. London: Routledge.

Sibley, D. (1999) 'Creating geographies of difference', in D. Massey, J. Allen and P. Sarre (eds), *Human geography today*. Cambridge: Polity Press, pp. 115–28.

Simpson, G.G. (1963) 'Historical science', in C.C. Albritten (ed.), *The fabric of geology*. Reading, MA: Addison–Wesley, pp. 24–48.

Sinclair, I. (1997) *Lights out for the territory: 9 excursions in the secret history of London*. London: Granta Books.

Slater, D. (1999) 'Situating geopolitical representations: inside/outside and the power of imperial interventions', in D. Massey, J. Allen and P. Sarre (eds), *Human geography today*. Cambridge: Polity Press, pp. 62–84.

Slater, D. (2000) 'Other domains of democratic theory: space, power and the politics of democratization', *Environment and Planning D: Society and Space*, vol. 20, pp. 255–76.

Smith, C. and Agar, J. (eds) (1998) *Making space for science: territorial themes in the shaping of knowledge*. Basingstoke: Macmillan.

Soja, E. (1989) *Postmodern geographies: the reassertion of space in critical social theory*. London: Verso.

Soja, E. (1996) *Thirdspace: journeys to Los Angeles and other real-and-imagined places*. Oxford: Blackwell.

Soustelle, J. (1956) *La vida cotidiana de los Aztecas en visperas de la conquista*. Mexico City: Fondo de Cultura Economica. (Originally published 1955 as *La vie quotidienne des Aztéques à la veille de la conquête espagnole*. Paris: Librairie Hachette.)

Spinoza, B. (1985) *Ethics*, in *The collected works of Spinoza* (trans. E. Curley). Princeton, NJ: Princeton University Press.

Spivak, G. (1985) 'The Rani of Sirmur', in F. Barker, P. Hulme, M. Iverson and D. Loxley (eds), *Europe and its Others*. Colchester: University of Essex Press, vol. 1, pp. 128–51.

Spivak, G. (1990) 'Poststructuralism, marginality, postcoloniality and value', in P. Collier and H. Geyer-Ryan (eds), *Literary Theory Today*. Ithaca, NY: Cornell University Press.

Staple, G. (1993) 'Telegeography and the explosion of place', *Telegeography, global traffic statistics and commentary*, pp. 49–56 (cited in Graham, 1998).

Stengers, I. (1997) *Power and invention: situating science*. Minneapolis and London: University of Minnesota Press.

Thrift, N. (1996) *Spatial formations*. London: Sage.

Thrift, N. (1999) 'The place of complexity', *Theory, Culture and Society*, vol. 16, no. 3, pp. 31–69.

Till, J. (2001) 'Eisenman's banana: review of Benjamin, 1999', *Radical Philosophy*, no. 108, pp. 48–50.

Townsend, R.F. (1992) *The Aztecs*. London: Thames and Hudson.

Tschumi, B. (1988) 'Parc de la Villette, Paris', *Architectural Design*, vol. 58, no. 3/4, pp. 32–9.

Tschumi, B. (2000a) 'Six concepts', in A. Read (ed.), *Architecturally speaking: practices of art, architecture and the everyday*. London: Routledge, pp. 155–76. (First published 1994 in *Architecture and disjunction*. Cambridge, MA: MIT Press; the page references are from the 2000 publication.)

Tschumi, B. (2000b) *Event cities 2*. Cambridge, MA: MIT Press.

Tuan, Y.F. (1977) *Space and place*. London: Arnold.

Tully, J. (1995) *Strange multiplicity: constitutionalism in an age of diversity*. Cambridge: Cambridge University Press.

Turner, B.S. (2002) 'Cosmopolitan virtue, globalization and patriotism', *Theory, Culture and Society*, vol. 19, nos. 1–2, pp. 45–63.

Urban Task Force (1999) *Towards an urban renaissance* (The Rogers Report). London: DETR.

Vaillant, C.G. (1950) *The Aztecs of Mexico*. Harmondsworth: Penguin.

van den Berg, C. (1997) 'Battle sites, mine dumps, and other spaces of perversity', in S. Golding (ed.), *The eight technologies of otherness*. London: Routledge, pp. 297–305.

Waddell, H. (1987) *The desert fathers*. London: Constable.

Wainwright, H. and Elliott, D. (1982) *The Lucas Plan: a new trade unionism in the making*. London: Allison and Busby.

Walker, R.B.J. (1993) *Inside/outside: international relations as political theory*. Cambridge: Cambridge University Press.

Walzer, M. (1995) 'Pleasures and costs of urbanity', in P. Kasinitz (ed.), *Metropolis: center and symbol of our times*. New York: New York University Press.

Wark, M. (1994) *Virtual geography: living with global media events*. Bloomington: Indiana University Press.

Watson, S. (1998) 'The new Bergsonism: discipline, subjectivity and freedom', *Radical Philosophy*, no. 92, pp. 6–16.

Weiss, L. (1998) *The myth of the powerless state*. Cambridge: Polity Press.

Whatmore, S. (1997) 'Dissecting the autonomous self: hybrid cartographies for a relational ethics', *Environment and Planning D: Society and Space*, vol. 15, pp. 37–53.

Whatmore, S. (1999) 'Hybrid geographies: rethinking the "human" in human geography', in D. Massey, J. Allen, and P. Sarre (eds), *Human geography today*. Cambridge: Polity Press, pp. 22–39.

Whatmore, S. and Hinchliffe, S. (2002/3) 'Living cities: making space for urban nature', *Soundings: a journal of politics and culture*, no. 22, pp. 37–50.

Wheeler, W. (1994) 'Nostalgia isn't nasty: the postmodernising of parliamentary democracy', in M. Perryman (ed.), *Altered states: postmodernism, politics, culture*. London: Lawrence and Wishart, pp. 94–109.

Wheeler, W. (1999) *A new modernity: change in science, literature and politics*. London: Lawrence and Wishart.

Whitehead, A.N. (1927/1985) *Symbolism: its meaning and effects*. New York: Fordham University Press.

Wigley, M. (1992) 'Untitled: the housing of gender', in B. Colomina (ed.), *Sexuality and space*. New York: Princeton Architectural Press, pp. 327–89.

Williams, R.S. Jnr (2000) 'The modern earth narrative: natural and human history of the earth', in R. Frodeman (ed.), *Earth matters: the earth sciences, philosophy, and the claims of community*. Upper Saddle River, NJ: Prentice Hall.

Wilmsen, E.N. (1989) *Land filled with flies: a political economy of the Kalahari*. Chicago and London: University of Chicago Press.

Wilson, E. (1991) *The sphinx in the city: urban life, the control of disorder, and women*. London: Virago Press.

Windley, B.F. (1977) *The evolving continents*. London: Wiley.

Wolf, E. (1982) *Europe and the people without history*. London: University of California Press.

Young, R. (1990) *White mythologies: writing history and the West*. London: Routledge.

Yuval-Davis, N. (1999) 'What is "transversal politics"?', *Soundings: a journal of politics and culture*, no. 12, pp. 94–8.

Zohar, D. (1997) *Rewiring the corporate brain: using the new science to rethink how we structure and lead organizations*. San Francisco: Berret–Koehler.

index

Index by Margaret Binns